METHODS IN MOLECULAR BIOLOGY™

Series Editor
John M. Walker
School of Life Sciences
University of Hertfordshire
Hatfield, Hertfordshire, AL10 9AB, UK

For further volumes:
http://www.springer.com/series/7651

Plant DNA Fingerprinting and Barcoding

Methods and Protocols

Edited by

Nikolaus J. Sucher

Department of Science, Technology, Engineering & Math (S.T.E.M.), Roxbury Community College, Roxbury Crossing, MA, USA

James R. Hennell

Centre for Complementary Medicine Research, School of Science and Health, University of Western Sydney, Penrith, NSW, Australia

Maria C. Carles

Department of Natural Sciences, Northern Essex Community College, Haverhill, MA, USA

 Humana Press

Editors
Nikolaus J. Sucher
Department of Science, Technology,
Engineering & Math (S.T.E.M.)
Roxbury Community College
Roxbury Crossing, MA, USA

James R. Hennell
Centre for Complementary Medicine Research
School of Science and Health
University of Western Sydney
Penrith, NSW, Australia

Maria C. Carles
Department of Natural Sciences
Northern Essex Community College
Haverhill, MA, USA

ISSN 1064-3745   e-ISSN 1940-6029
ISBN 978-1-61779-608-1   e-ISBN 978-1-61779-609-8
DOI 10.1007/978-1-61779-609-8
Springer New York Dordrecht Heidelberg London

Library of Congress Control Number: 2012932855

Printed on acid-free paper

Humana Press is part of Springer Science+Business Media (www.springer.com)

Preface

Over the last 25 years or so, molecular cloning and DNA-based analysis have become part of every molecular life science laboratory. The rapid adoption of DNA-based techniques has been facilitated by the introduction of the polymerase chain reaction (PCR) over 20 years ago, which has made cloning and characterization of DNA quick and relatively simple. PCR is part of virtually every variation of the plethora of approaches used for DNA fingerprinting today. Looking back at the rapid development and change in methodology used for DNA fingerprinting between the publication of the first (1) and second (2) editions of their book entitled "DNA Fingerprinting in Plants," the authors wondered what a similar book might look like in 2015. This volume may serve as a waypoint. A significant part of this volume deals with DNA barcoding, a term that did not even appear in the previously published compendium on DNA fingerprinting (2).

We have tailored this volume principally for those who seek to augment their current methods of plant analysis and quality control using genome-based approaches. We aim to bring together the different currently available genome-based techniques into one repository. Included also are several discussions around the broader issues of genome-based approaches in order to provide a sound understanding of the principles of these methods so that this volume may be useful to others involved in different plant sciences.

This book contains detailed protocols for the preparation of plant genomic DNA, fingerprinting of plants for the detection of intraspecies variations as well as the use of DNA barcoding. Methods for the bioinformatics analysis of data are also included.

Roxbury Crossing, MA, USA
Campbelltown, NSW, Australia
Haverhill, MA, USA

Nikolaus J. Sucher
James R. Hennell
Maria C. Carles

References

1. Weising K (1995) *DNA fingerprinting in plants and fungi* (CRC Press, Boca Raton) p 322.

2. Weising K (2005) *DNA fingerprinting in plants: principles, methods, and applications* (Taylor & Francis Group, Boca Raton, FL) 2nd Ed p 444.

Contents

Contributors

MARY V. ASHLEY • *Department of Biological Sciences, University of Illinois at Chicago, Chicago, IL, USA*

BERNARD R. BAUM • *Eastern Cereal and Oilseed Research Centre, Agriculture & Agri-Food Canada, Ottawa, ON, Canada*

CINZIA MARGHERITA BERTEA • *Plant Physiology Unit, Department of Plant Biology, University of Turin, Turin, Italy*

MARIA C. CARLES • *Department of Natural Sciences, Northern Essex Community College, Haverhill, MA, USA*

PREETI CHAVAN • *Interactive Research School for Health Affairs, Bharati Vidyapeeth University, Pune, India*

ROBYN S. COWAN • *Jodrell Laboratory, Royal Botanic Gardens, Richmond, Surrey, UK*

PAUL M. D'AGOSTINO • *School of Biomedical and Health Sciences, School of Science and Health, University of Western Sydney, Penrith, NSW, Australia*

PATRICIA L. DIAZ • *Centre for Complementary Medicine Research, School of Science and Health, University of Western Sydney, Penrith, NSW, Australia*

EUI JEONG DOH • *Department of Biological sciences, Konkuk University, Seoul, Republic of Korea*

MICHAEL F. FAY • *Jodrell Laboratory, Royal Botanic Gardens, Richmond, Surrey, UK*

GIORGIO GNAVI • *Plant Physiology Unit, Department of Plant Biology, University of Turin, Turin, Italy*

SCOTT J. HARPER ⁿ *Citrus Research and Education Center, Laboratory University of Florida, Lake Alfred, FL, USA*

JAMES R. HENNELL • *Centre for Complementary Medicine Research, School of Science and Health, University of Western Sydney, Penrith, NSW, Australia*

TRIPTA JHANG • *Department of Genetic Resource Management, Division of Genetics and Plant Breeding, Central Institute of Medicinal and Aromatic Plants, Lucknow, India*

KALPANA JOSHI • *Department of Biotechnology, Sinhgad College of Engineering, Pune, India*

CHEANG S. KHOO • *Centre for Complementary Medicine Research, School of Science and Health, University of Western Sydney, Penrith, NSW, Australia*

MIN-KYEOUNG KIM • *Department of Oriental Medicinal Materials and Processing, College of Life Science, Kyung Hee University, Suwon, South Korea*

OK-RAN LEE • *Department of Oriental Medicinal Materials and Processing, College of Life Science, Kyung Hee University, Suwon, South Korea*

SAMIUELA LEE • *Centre for Complementary Medicine Research, School of Science and Health, University of Western Sydney, Penrith, NSW, Australia*

JIAN-FENG LI • *Department of Molecular Biology, Massachusetts General Hospital, Boston, MA, USA*

SEUNG-EUN OH • *Department of Biological sciences, Konkuk University, Seoul, South Korea*

OVIDIU PAUN • *Department for Systematic and Evolutionary Botany, University of Vienna, Vienna, Austria*

MARYAM SARWAT • *Pharmaceutical Biotechnology, Amity Institute of Pharmacy, Amity University, Uttar Pradesh, Noida, India*

PETER SCHÖNSWETTER • *Institute of Botany, University of Innsbruck, Innsbruck, Austria*

AJIT KUMAR SHASANY • *Department of Plant Biotechnology, Division of Biotechnology, Central Institute of Medicinal and Aromatic Plants, Lucknow, India*

JEN SHEEN • *Department of Genetics, Harvard Medical School, Boston, MA, USA*

NIKOLAUS J. SUCHER • *Department of Science, Technology, Engineering & Math (S.T.E.M.), Roxbury Community College, Roxbury Crossing, MA, USA*

LISA I. WARD • *Plant Health and Environment Laboratory, Investigation and Diagnostic Centre, MAF Biosecurity New Zealand, Wellington, New Zealand*

DEOK-CHUN YANG • *Department of Oriental Medicinal Materials and Processing, College of Life Science, Kyung Hee University, Suwon, South Korea*

DAVID N. ZAYA • *Department of Biological Sciences, University of Illinois at Chicago, Chicago, IL, USA*

Chapter 1

A Taxonomist's View on Genomic Authentication

Bernard R. Baum

Abstract

A brief history of taxonomy, for the most part plant oriented, is provided, which demonstrates the use of morphology early on, through the stages when different technologies became available at different times until the present use of genomic tools. Genomic authentication facilitates with greater precision than ever before the identification of an organism or part thereof. In this chapter I made an attempt to stress that, in general, but more so for genomic authentication, the use of the variation inherent in taxa down to the lowest level of the hierarchy of classification needs to be used to achieve a high degree of correct authentication.

Key words: Taxonomy, Flowering plants, Taxonomic botany, Genomic authentication, *Echinacea purpurea*

1. What Are Taxonomists, What Do They Practice, for What Purpose and What Tools Do They Use?

1.1. Background

Taxonomists practice taxonomy, the science of delimiting and classifying objects, artifacts, languages and sounds, organisms, diseases, etc. We will, usually but not always, restrict our subject to taxonomists of flowering plants. Taxonomy is an essential preoccupation of many aspects in daily human life. Taxonomy has its roots in prehistoric times when humans were hunters and food gatherers. In those early times, humans needed to distinguish between beneficial organisms and non-beneficial ones for their survival, such as non-poisonous and poisonous plants. Briefly, humans had to differentiate between the good and the bad for survival. With time, the need to differentiate between the different kinds of animals and plants developed along with signs and words to labels them. Early stages of classification and identification were then born along with nomenclature albeit informal. These naming systems are called "folk taxonomies" by ethnobotanists, which in contrast to scientific taxonomy are based on social or cultural traditions (1).

Nikolaus J. Sucher et al. (eds.), *Plant DNA Fingerprinting and Barcoding: Methods and Protocols*, Methods in Molecular Biology, vol. 862, DOI 10.1007/978-1-61779-609-8_1, © Springer Science+Business Media, LLC 2012

Folk taxonomies often do not correspond closely to scientific taxonomy. For instance, within a sample of 200 native names of plants in a Tzeltal-speaking community, Chiapas, Mexico, 41% of the names covered more than one scientific species, whereas 25% referred only partially to a scientific species (2). In Lake Malawi/Nyasa in East Africa, 536 local name usages of fish species observed in three localities were found corresponding with Tonga, Tumbuka, and Nyanja/Yao languages, suggesting that the locals identify fish that correspond to scientific taxonomy (3). This and the following example stand in contrast to another example from the folk taxonomy of native people in New Guinea who recognized 136 of the 137 taxonomic bird species living in the mountains of this country (4).

Folk taxonomy is a very large subject and its anthropological aspects are outside the subject of this book. However, it has formed the basis for many of the names of genera and species in scientific taxonomy.

Early civilizations, such as the Greeks, developed systems of classes along with descriptions and identification means based on visual features, such as Theophrastus (5) (370–285 BC) who is the first person in the European tradition of plant taxonomy concerned with the medical qualities of 480 plants and in identifying their essential features. Following him the Greek herbalist Dioscorides (6) described 600 species of oil, spices, grain cereals, herbs, and others. Dioscorides' book was used during medieval times when taxonomy did not progress until an increased interest in medicinal plants arose as attested by an increase in a number of herbal publications, such as De Plantis by Caesalpinus (7). It was, however, not easy to recognize the increasing number of plants of interest and to memorize their complex names. An example is "Verbena foliis multifido-laciniatis, spicis filiformibus" described by Dalibard (8) also known as "Verbena communis, flore caeruleo" described by Bauhin (9) and known as the Common Vervain (*Verbena officinalis* L.). It was used in herbalism, a traditional medicinal or folk medicine practice based on the use of plants and plant extracts. Herbalism is also known as botanical medicine, medical herbalism, herbal medicine, herbology, and phytotherapy. The common Vervain was used as herbal tea, as a sex steroid, to treat nervousness and insomnia.

Carl Linnaeus (born 1707, died 1778), the Father of taxonomy, developed a system for naming, ranking, and classifying organisms which is still in use today. Linnaeus' ideas on classification have influenced generations of biologists. Linnaeus (10) described about 6,000 plant species and changed the then traditional cumbersome naming into a much simpler binomial nomenclature, commonly and currently used. Binomial nomenclature treats species as subordinate to genera, as in the above example *Verbena* the genus containing several species among them *V. officinalis.*

Today, taxonomists write local and regional floras with identification keys. Taxonomists use various tools to obtain data such as morphology, anatomy, chemistry, cytology…molecular methods and bioinformatics to refine classifications (see the Tree of Life, http://tolweb.org/tree/) to authenticate identification with increased precision (see barcoding below) and examining fragmentary materials for the identifying, as in the case of forensics or in the case of an ingested poisonous organism for instance.

1.2. Purpose

It may become apparent from the brief background above that taxonomy in general was carried out for utilitarian purpose. But there is another reason; which is curiosity or the pursuit of knowledge. Knowledge provides the understanding of the world in which mankind lives and this knowledge carries the potential for utility and conservation. Taxonomy may be divided into several activities which are often intertwined and yielding several end products, namely classification, identification, nomenclature, and, not least, the study of taxonomic principles and theory.

1.3. Tools Used

I will focus particularly on taxonomic botany. Tools to obtain the first three end products were developed from the beginnings of taxonomy and with time new tools were added. Different methods were developed often based on newly available technologies and based on different philosophies or novel thinking to obtain the same end product. The first developed tool was simply the visual description, i.e., morphology or habit. With technological advances, i.e., lenses and the microscope, micromorphology and anatomy became added as tools. Over time taxonomists have added many tools to their repertoire in order to achieve better taxonomies. Taxonomy ought to be regarded quite rightly as an unending synthesis (11). Before continuing to enumerate the tools one should make a distinction between tools for data acquisition and tools for data analysis.

A wide array of data acquisition has increasingly become available to taxonomists (Table 1). Concurrently new methods of data analysis were developed (Table 2), among them some specific to the data kinds. For instance, jModelTest (12), being a method with which one estimates the evolutionary model with its parameter estimates from the data, was developed after DNA sequences became relatively easily accessible and after some different phylogenetic models were invented, the first of the models being Jukes and Cantor (13).

2. What Is Genomic Authentication?

To prove authenticity means acceptance or belief based on fact. For instance a painting is authenticated by verification of its painter's original signature. Notwithstanding a signature it could be

Table 1
**Methods of data acquisition that became available
to taxonomists over time and tools**

Discipline	Tool	Material
Gross morphology	Visual	Herbarium specimens (HS)
Morphology	Dissecting microscope	Live specimens (LS) + HS
Micromorphology	Microscope (M)	LS + HS
Anatomy	M	LS + HS
Micromorphology	Electron microscope (EM)	LS + HS
Palynology	M + EM	LS + HS + fossil specimens (FS)
Cytology	M and EM	LS + HS
Cytogenetics	Crossing experiments	LS
Chemistry	Laboratory, chemical reactions	LS
Phytochemistry	HPLC	LS
Protein chemistry	Electrophoresis (E)	LS
Serology	Serological reactions	LS
Enzymes	E	LS + seed material
DNA	Hybridization	LS
DNA	E	LS + HS
Protein	E	LS + HS
DNA	Sequence (S)	LS + HS + FS
Protein	S	LS

Transition was from the visual to microscopical, to chemical, first using micromolecules then macromolecules

authenticated by a chemical analysis of the painting followed by a comparison with paintings by the same artist. Similarly, a particular plant part can be authenticated by the verification of one or more of its features, such as morphology, micromorphology, anatomy, etc. including of course its DNA characteristics (Table 1).

What exactly do we mean by features and especially by DNA characteristics? To answer this question depends on what is our objective. Is our objective identification to a particular category such as family, genus, species, subspecies or strain? Or do we try to identify a fragment from its original specimen? Generally one would need tools for markers discovery, validation and verification. In other

Table 2
Methods of data analysis and their products that became available to taxonomists over time

Method of analysis	Product	Data
Neurological, i.e., manual	Identification keys	Descriptive characters
Computer programs	Identification keys, online identification	Descriptive characters, morphometric characters
Neurological	Establishing classes, groupings	Descriptive characters, mainly qualitative
Numerical taxonomy	Establishing classes, groupings	Morphometrics, quantitative and qualitative characters
Word and string matching algorithms	DNA sequence alignment	DNA sequences and amino acid sequences
Phylogenetic analysis	Establishing classes, groupings, gene trees	Discrete characters: qualitative, sometimes quantitative, DNA sequences, amino acid sequences
Population genetic analysis	Differentiation and relationship among populations, gene flow	Morphological characters, allelic data, isoenzyme data, DNA sequences
Manual	Distribution maps	Collection data, including herbarium label data
Computerized, GIS, etc.	Distribution maps	Collection data, including herbarium label data
Co-phylogenetic analysis, mainly computerized	Host–parasite relations	Discrete characters: qualitative, sometimes quantitative, DNA sequences, amino acid sequences

Transition from using the mind alone to computerized analysis and from simple methodology to elaborate methodologies

words, one needs to establish a reference system or a database containing the description of the features and their variation of the recognized categories, these features that delimit the categories and these features useful for identification. Such a reference system or a database would require to be updated on a regular basis.

Genomic authentication is based on biomarker development. Biomarkers are molecular phenotypes in the form of a pattern, like a fingerprint that is used to recognize an individual or a group of individuals in different taxonomic categories (see above), such as a species. The development of biomarkers entails a discovery phase followed by validation and then by verification. Biomarkers are of different kinds and different approaches and methods. Biomarkers are increasingly developed as new and efficient methods and techniques are being invented.

Earlier genomic authentication methods were based on DNA hybridization. The "restriction fragment length polymorphism" (RFLP) was one of the more advanced early techniques and thus became very popular for a while. RFLP was based on the method of Southern (14). With the advent of the polymerase chain reaction (PCR) which amplifies specific regions of interest in the genome using nucleotide primers, a technique invented by Mullis et al. (15) and Saiki et al. (16), explosive developments of methods with widespread applications emerged including nucleotide sequencing (17). A few examples of the most practiced array of marker systems is: random amplified polymorphic DNA (RAPD), simple sequence repeat polymorphism (SSR), cleavable amplified polymorphic sequences (CAPS), amplified restricted fragment length polymorphism (AFLP™), inter-repeat amplification (IRA), which were described compared and evaluated (18) and for a related review with emphasis on AFLP see ref. (19). The authentication of several of these methods as applied to medicinal plants has recently been reviewed including molecular technical details by Techen et al. (20) who concluded that there is no single or superior method to assure 100% authentication and that a direction to reach this goal can be obtained by a variety of techniques that include traditional ones, i.e., non-molecular ones.

Recently a new class of biomarkers has become the method of choice for genomic identification and authentication. Single nucleotide polymorphisms (SNPs) are the most frequent kind of polymorphism in genomes, which is advantageous over previous methods of variation assessment. This is because genetic variation is ubiquitous among and especially within groups of organisms, a fact that needs to be taken into consideration in the development of biomarkers for authentication especially at and below the species level. SNPs can be divided into two kinds, those occurring in non-coding regions and those in coding regions. For instance, in rice there is one SNP per 170 bp throughout its genome (21), in cotton wood (*Populus trichocarpa* Hooker) one SNP per 64 bp was found in non-coding regions and one SNP per 229 bp in coding regions (22). Millions of SNPs are indexed in the National Center for Biotechnology Information's dbSNP database, covering organisms from *Anopheles gambiae* to *Zea mays* (23). Each taxonomic category can be characterized by the ensemble of its SNPs which then becomes available for precise identification. The advantage of SNPs as a tool for genotyping (24) and detecting variation among species and below became possible and practical with the advent of high-throughput sequencing methods and detection and validation. Furthermore, methods that do not require electrophoresis for SNP genotyping become increasingly available, for instance the method of Hirotsu et al. (25). However, it will take the next generation sequencing platforms from such companies as Illumina Inc., Helicos Biosciences Corp., and Applied Biosystems™, a division of Life Technologies Corp., to enable generating low cost high throughput large volume of sequence data to enable the production of gene chips for the identification of genotypes, a key tool in the future of genomic authentication. An example of a test of whether one of these methods is suitable for use in polyploid wheats using the SNPlex™ genotyping system of Applied Biosystems was carried out on a limited sample by Bérard et al. (26). In the meantime and until these methods become more affordable, several attempts were made with traditional genomic tools.

All of the above genotyping methods and tools developed thereof are aimed at providing variation data needed for authentication based on variation. DNA barcoding methods on the other hand and although genomically based (27–29), are tools being developed on the belief that a very small stretch of the genome, just several base pairs or short orthologous sequences from several specific areas of the plant genome, are enough to authenticate species. An international initiative aimed at developing DNA standard for the identification of biological is known as "The Consortium for the Barcode of Life" (CBOL) http://www.barcodeoflife.org/. CBOL is definitely dedicated to the identification or authentication of the category of species; see ref. also (30).

3. Limitation of Genomic Authentication

The most important aspect of genomic authentication is the integration of genomic information with taxonomic information. This is related to the species problem, i.e., on what grounds is a decision made that a group of organisms constitutes a species. Although the species is accepted as the basic unit of classification in the hierarchy of organisms, species concepts abound. The different concepts also depend on philosophical issues such as realism and nominalism. The literature on this subject is enormous and it is not germane here to delve into the issue, except to mention that a chapter by Mayr (31) is a worthwhile historical introduction. Species as entities, whether defined in terms of morphological discontinuity or restriction of gene exchange, they are unique in that they are non-arbitrary as to both inclusion and exclusion (32). These entities can differ from each other by their genome and observed by various features (characters) as in Table 1. The majority of plant species were established first and foremost on the basis of morphology. Categories below the species level were also, but not only, established on morphology and quite often variant categories were distinguished by other features, such as chemistry ones (chemovars) and other ones (Table 1). The distinction of such categories is necessary for instance among nutraceuticals, medicinal plants, and cultivars in crops, ornamentals, and forestry. Thus the delimitation of species and of groups below the species requires data from different sources including morphology, cytology, anatomy, etc., and certainly multiple molecular markers (33).

Genomic authentication may fail for a number of reasons which may be technical, as for instance gel-based techniques which are often prone to laboratory conditions. DNA probes in RFLP are inherently limited because the mechanics of hybridization is restricted mainly to the two ends of the probes and dependence to the assay level of stringency. One of the main reasons for the failure of genomic authentication is the inherent variation within the group category, usually species and its variants, or that the group is not fully understood and its species or groups are ill-defined. Another reason is the limitation of the method of authentication or worse when the group in question is poorly understood, complicated taxonomically, or extremely variable; that DNA barcoding in wild potatoes fails (34) and similar example from insects has been reported (35). To achieve accuracy in identification it has been suggested that longer DNA fragments than those currently suggested for barcoding are desirable (36), that other sources need to be explored in the genome especially for barcoding (37), and that adequate sampling for DNA sequence diversity needs to be taken from the range of distribution of the groups in the study (33).

One of the biggest problems of genome authentication is due to mixtures, adulterations, and hybrids. A sample may appear to be morphologically homogeneous to the eye and therefore needs to be sorted out by other means first, such as the second to seventh in Table 1 from the top. Hybrids are trickier. After the initial triage genomic authentication can be used. In reality however, one ought not to expect to base genome authentication on uniformity or on homogeneity of the sample even at the genomic level. The reality is that variation is a natural phenomenon as shown in our experience with genomic identification of barley (*Hordeum*) cultivars (see below). Basically, it is the ensemble of the variation that was found to be typical to the taxa, barley cultivars in this case. This typical genomic variation to each taxon needs to be considered as the basis for authentication.

4. Example of Genomic Authentication of a Plant, e.g., Echinacea

In herbal plants used for pharmaceutical purposes two aspects are needed for quality assurance, one being the authentication of the species and the other is the authentication of the concentration of the particular phytochemical(s). *Echinacea* is endemic to North

America and one species, *Echinacea purpurea*, is cultivated in several countries.

For the first aspect we carried out a taxonomic revision of the genus from herbarium collections and by collecting sample material (seed and live plants) from 58 native populations throughout the area of its distribution, i.e., United States and Canada, and by assembling cultivated material from growers. We then carried out a taxonomic analysis of the material to delimit and circumscribe the species, to produce an identification key to the species and to infer the species evolutionary relationships (38). The identification key is useful for the identification of species and varieties based on morphology.

Concurrent to the morphological study, a genomic investigation was carried out on the material from the 58 native populations. At the time AFLP was the technique of choice based on genome coverage and cost considerations. The outcome of the AFLP study was that from a total of 435 individual plants investigated the resulting polymorphism was sufficient to distinguish each plant; however, no species-specific and no variety-specific fingerprints were found (39). Identification at the species level was achieved with a minimum of ten AFLP fragments with greater than 82% correct classification using a classification function coefficient generated by a canonical analysis of the fragment scores data subsequently tested by cross-validation. A linear discriminant function coefficient based on the ten variables was provided to enable genome authentication to species only ((39): Table 11).

It is noteworthy that AFLP bands in an AFLP profile (a gel) are only identical in their mobility. We tested the sequence identity of the monomorphic and polymorphic bands by sequencing several bands of each (40). The monomorphic bands exhibited above 90% sequence identity among DNA clones within samples; whereas within variety they ranged from 82 to 94%; within species from 75 to 98%; and within the genus 58%. In contrast the polymorphic bands exhibited 51–100% within sample, 33–100% within variety, 23–45% within species, and as low as 1.25% within the genus. We demonstrated that co-migrating bands in AFLP cannot be considered homologous and concluded that they should be considered phenotypically identical but not genetically.

Regarding the second aspect, authentication of presence of phytochemicals by genomic analysis, we tested whether the level of dodeca-2E, 4E, 8Z, 10E/Z-tetraeonoic acid isobutyl amide (1, 2) and cichoric acid in some lines and clonal materials can be predicted by DNA fingerprints, specifically AFLP (41). The results showed that although there was a weak correlation between the AFLP fingerprints and the phytochemical content, there was a strong association (trend) between the two sets.

5. Example of Genomic Authentication of a Crop Plant: Cultivar Identification, e.g., Six-Rowed Spring Malting Barley Cultivars

Malting barleys are important to the brewing industry. About 9,000 cultivars (varieties) have been developed in the single cultivated barley species (*Hordeum vulgare* L.) of which many earlier ones are not in use anymore. Currently it has become very difficult to identify one cultivar from the other by morphological or biochemical characters alone. We have shown (42) that a set of 15 SNPs was necessary and sufficient to identity the 17 six-rowed spring malting barleys then grown in Canada. The 15 SNPs were selected from a large set that were obtained by sequencing and data mining. An important observation was that even at this low level of the hierarchy of life, the cultivar, we found a varying number of SNP variants per cultivar. For instance from the 17 SNPs used, cultivar "Duel" had one SNP but cultivar "Bonanza" in contrast had four and "Foster" had five. To authenticate a cultivar using a microchip one would need to use all the different characteristic SNPs necessary in combination for the identification.

6. Conclusion

Taxonomy is an open-ended science of classification and identification; it is constantly renovating itself. Taxonomy entails analysis and synthesis. When a new technology becomes available, taxonomy is quick to adopt it for its never-ending quest of producing more refined and more precise classification and identification. The more recent tools including genomics enable more precise authentication on the basis of a plurality of data taken from the variation inherent in the taxa. In this respect, DNA sequence data in various forms, such as SNPs, facilitate higher precision authentication.

Acknowledgments

I am indebted to my colleague John Thor Arnason, University of Ottawa, for his comments and suggestions to improve the manuscript. I am grateful to NJ Sucher, University of Western Sydney, Penrith South, Australia, for the invitation to write this chapter.

References

1. Raven PH, Berlin B, Breedlove DE (1971) The origins of taxonomy. Science 174: 1210–1213. DOI: 10.1126/science.174.4015.1210.

2. Berlin B, Breedlove DE, Raven PH (1966) Folk taxonomies and biological classification. Science 154: 273–275.

3. Ambali A, Kabwazi H, Malekano L, Mwale G, Chimwaza D, Ingainga J, Makimoto N, Nakayama S, Yuma M, Kada Y (2001) Relationship between local and scientific names of fishes in Lake Malawi/Nyasa. African Study Monographs 22: 123–154.

4. Mayr E (1941 The origin and history of the bird fauna of Polynesia. *Proceedings of the 6th Pacific Scientific Congress* 4:197–216.

5. Theophrastus, E. 364 BC? De Historia Plantarum.[also a book printed 1644 in Amsterdam under the title Theophrasti Eresii. De historia plantarum libri decem Graece et Latine … totum opus … cum notis …item rariorum plantarum iconibus illustravit Joannes Bodaeus a Stapel],,

6. Dioscorides, PA. 65 AD [2000]. De Materia Medica being an herbal with many other medicinal materials written in Greek in the first century of the common era. *A new indexed version in modern English* by TA Osbaldeston and RPA Wood. IBIDIS press, Johannesburg.

7. Caesalpinus 1583) De Plantis.

8. Dalibard, T.F (1749) Florae Parisiensis Prodromus. Paris.

9. Bauhin, C (1623) Theatri botanici. Basel.

10. Linnaeus C (1753) Species plantarum :exhibentes plantas rite cognitas, ad genera relatas, cum differentiis specificis, nominibus trivialibus, synonymis selectis, locis natalibus, secundum systema sexuale digestas. Stockholm.

11. Constance L (1964) Systematic botany – an unending synthesis. Taxon 257–273.

12. Posada D (2008) jModelTest: Phylogenetic model averaging. Mol. Biol. Evol. 25: 1253–1256.

13. Jukes TH, Cantor CR (1969) Evolution of protein molecules pp 21–132 In Munro HN (ed) Mammalian protein metabolism. Academic Press, New York.

14. Southern EM (1975) Detection of specific sequences among DNA fragments separated by gel electrophoresis. Jour. Mol. Biol. 98: 503–517.

15. Mullis KB, Faloona FA, Scharf SJ, Saiki RK, Horn GT, Erlich HA (1986) Specific enzymatic amplification of DNA in vitro: the polymerase chain reaction. Cold Spring Harbor Symp. Quant. Biol. 51: 263–273.

16. Saiki RK, Scharf SJ, Faloona FA, Mullis KB, Horn GT, Erlich HA, Arnheim N (1985) Enzymatic amplification of β-globin genome sequences and restriction site analysis for diagnosis of sickle cell anemia. Science 230: 1350–1354.

17. White BA (editor). PCR protocols: current methods and applications. Humana Press, Totowa NJ.

18. Rafalski JA, Vogel JM, Morgante M, Powell W, Andre C, Tingey SV (1996 Generating and using DNA markers in plants. In: Birren B, Lai E (eds) Nonmammalian genomic analysis. A practical guide. Academic Press, San Diego, pp 75–134.

19. Mueller UG, Wolfenbarger LL (1999) AFLP genotyping and fingerprinting. TREE 14: 389–394.

20. Techen N, Crockett SL, Khan IA, Scheffler BE (2004) Authentication of medicinal plants using molecular biology techniques to compliment conventional methods. Current Medicinal Chemistry 11: 1391–1401.

21. Yu J, Hu S, Wang J, Wong GK, Li S, Liu B, Deng Y, Dai L, Zhou Y, Zhang X, Cao M, Liu J, Sun J, Tang J, Chen Y, Huang X, Lin W, Ye C, Tong W, Cong L, Geng J, Han Y, Li L, Li W, Hu G, Huang X, Li W, Li J, Liu Z, Li L, Liu J, Qi Q, Liu J, Li L, Li T, Wang X, Lu H, Wu T, Zhu M, Ni P, Han H, Dong W, Ren X, Feng X, Cui P, Li X, Wang H, Xu X, Zhai W, Xu Z, Zhang J, He S, Zhang J, Xu J, Zhang K, Zheng X, Dong J, Zeng W, Tao L, Ye J, Tan J, Ren X, Chen X, He J, Liu D, Tian W, Tian C, Xia H, Bao Q, Li G, Gao H, Cao T, Wang J, Zhao W, Li P, Chen W, Wang X, Zhang Y, Hu J, Wang J, Liu S, Yang J, Zhang G, Xiong Y, Li Z, Mao L, Zhou C, Zhu Z, Chen R, Hao B, Zheng W, Chen S, Guo W, Li G, Liu S, Tao M, Wang J, Zhu L, Yuan L, Yang H: 2002. A draft sequence of the rice genome (Oryza sativa L. ssp indica). Science 296:79–92.

22. Gilchrist EJ, Haughn GW, Ying CC, Otto SP, Zhuang J, Cheung D, Hamburger B, Aboutorabi F, Kalynyak T, Johnson L, Bohlmann J, Ellis BE, Douglas CJ, Cronk QCB. (2006) Use of Ecotilling as an efficient SNP discovery tool to survey genetic variation in wild populations of Populus trichocarpa. Molecular Ecology 15: 1367–1378.

23. Perkel J (2008) SNP genotyping: six technologies that keyed a revolution. Nature Methods 5: 447–453.

24. Henry RJ (ed) 2008. Plant genotyping II. SNP technology. CAB International, Oxfordshire.

25. Hirotsu N, Murakami N, Kashiwagi T, Ujiie K, Ishimanu K (2010) Protocol: a simple gel-free

method for SNP genotyping using allele-specific primers in rice and other plant species. Plant Methods 6: 12, doi:10.1186/1746-4811-6-12.

26. Bérard A, Le Paslier MC, Dardevet M, Exbrayat-Vinson F, Bonnin I, Cenci A, Haudry A, Brunel D, Ravel C (2009) Hugh-throughput single nucleotide polymorphism genotyping in wheat (Triticum spp.). Plant Biotechnology Jour. 7: 364–374.

27. Kress WJ, Wurdack KJ, Zimmer EA, Weigt LA, Janzen DH (2005) Use of DNA barcodes to identify flowering plants. PNAS 102: 8369–8374.

28. Chase MW, Cowan RS, Hollingsworth PM, van den Berg C, Madriñán S, Petersen G, Seberg O, Jørgsensen T, Cameron KM, Carine M, Pedersen N, Hedderson TAJ, Conrad F, Salazar GA, Richardson JE, Hollingsworth ML, Barraclough TG, Kelly L, Wilkinson M (2007) A proposal for a standardised protocol to barcode all land plants. Taxon 56: 295–299.

29. Erickson DL, Spouge J, Alissa R, Weigt LA, Kress JW (2008) DNA barcoding in land plants: developing standards to quantify and maximize success. Taxon 57: 1304–1316.

30. Ratnasingham S, Hebert PDN (2007) BOLD: the barcode of life data system (www.barcodinglife.org). Molecular Ecology Notes doi: 10.1111/j.1471-8286.2006.01678.x

31. Mayr E (1982) Chapter 6 Microtaxonomy, the science of species. In Mayr E (1982) The growth of biological thought. Harvard Univ. Press.

32. Davis PH, Heywood VH (1963) Principles of angiosperm taxonomy. D. Van Norstrand, Princeton.

33. Funk DJ, Omland KE (2003) Species level paraphyly and polyphyly: frequency, causes and consequences, with insights from animal mitochondrial DNA. Annual Review Ecol. Evol. Syst. 34: 397–423.

34. Spooner DM (2009) DNA barcoding will frequently fail in complicated groups: an example in wild potatoes. American Jour. Bot. 96: 1177–1189.

35. Elias M, Hill RL, Willmot KR, Dasmahapatra KK, Brower AWZ, Mallet J, Jiggins CD (2007) Limited performance of DNA barcoding in a diverse community of tropical butterflies. Proc. Royal Soc. B. 274: 2881–2889) doi: 10.1098/rspb.2007.1035

36. Roe AD, Sperling FAH (2006) Patterns of evolution of mitochondrial cytochrome c oxidase I and II DNA and implications for DNA barcoding. Molecular Phylogenetics and Evolution, 44: 325–345.

37. Devey DS, Chase MW, Clarkson JJ (2009) A stuttering start to plant DNA barcoding: microsatellites present a previously overlooked problem in non-coding plastid regions. Taxon 58: 7–15.

38. Binns SE, Baum BR, Arnason JT (2002) A taxonomic revision of *Echinacea* (Asteraceae: Heliantheae). Systematic Botany 27: 610–632.

39. Mechanda SM, Baum BR, Johnson DA, Arnason JT (2004a. Analysis of diversity of natural populations and commercial lines of *Echinacea* using AFLP. Can. J. Bot. 82: 461–484.

40. Mechanda SM, Baum BR, Johnson DA, Arnason JT (2004b. Sequence assessment of comigrating AFLP™ bands in *Echinacea* -? Implications for comparative biological studies. Genome 47: 15–25.

41. Baum BR, Mechanda SM, Lievsey JF, Binns SE, Arnason JT (2001) Predicting quantitative phytochemical markers in single *Echinacea* plants or clones from their DNA fingerprints. Phytochemistry 56: 543–549.

42. Baum BR, Johnson DA, Soleimani VD (2008) Six rowed spring malting cultivars can be identified by single nucleotide polymorphisms (SNP) as detected by allele-specific PCR. In: Groendijk-Wilders N, Alexander C, van den Berg RG, Hetterscheid WLA (eds). Proc. Vth Int. Symp. Taxonomy Cult. Plants. Acta Horticulturae 799: 71–80.

Chapter 2

DNA Fingerprinting, DNA Barcoding, and Next Generation Sequencing Technology in Plants

Nikolaus J. Sucher, James R. Hennell, and Maria C. Carles

Abstract

DNA fingerprinting of plants has become an invaluable tool in forensic, scientific, and industrial laboratories all over the world. PCR has become part of virtually every variation of the plethora of approaches used for DNA fingerprinting today. DNA sequencing is increasingly used either in combination with or as a replacement for traditional DNA fingerprinting techniques. A prime example is the use of short, standardized regions of the genome as taxon barcodes for biological identification of plants. Rapid advances in "next generation sequencing" (NGS) technology are driving down the cost of sequencing and bringing large-scale sequencing projects into the reach of individual investigators. We present an overview of recent publications that demonstrate the use of "NGS" technology for DNA fingerprinting and DNA barcoding applications.

Key words: DNA fingerprinting, DNA sequencing, DNA barcoding, Next generation sequencing, Plant transcriptomics, Chloroplast genome

1. Introduction

Two articles that were published in the scientific journal *Nature* 105 years apart first demonstrated the use of fingerprints (1) and DNA "fingerprints" (2) for individual-specific identification of humans. While it took 25 years after the article in *Nature* from 1888 and the extensive work published in a monograph by Francis Galton (3) before fingerprints were first accepted as evidence in court in 1905, DNA fingerprints were accepted by the judicial system within a couple of years after first publication of the technique. Since then, DNA fingerprinting has become a universal tool for the characterization of genetic differences and relatedness of individuals even thousands of years after their death (4).

Nikolaus J. Sucher et al. (eds.), *Plant DNA Fingerprinting and Barcoding: Methods and Protocols*, Methods in Molecular Biology, vol. 862, DOI 10.1007/978-1-61779-609-8_2, © Springer Science+Business Media, LLC 2012

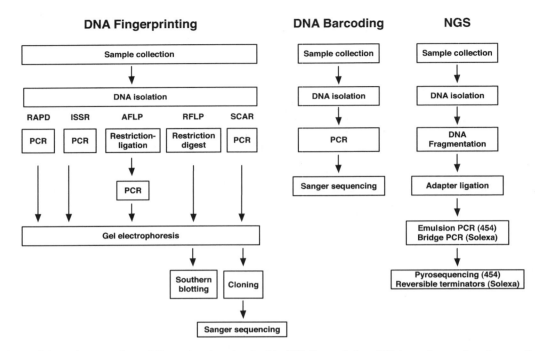

Fig. 1. Schematic comparison of the major steps involved in DNA fingerprinting, DNA barcoding, and next generation sequencing. *RAPD* random amplified polymorphic DNA (Chapter 6), *ISSR* inter simple sequence repeat (Chapter 9), *AFLP* amplified fragment length polymorphism (Chapter 7), *RFLP* restriction fragment length polymorphism (Chapter 8), *SCAR* sequence characterized amplified region (Chapter 10), DNA barcoding (Chapters 3 and 11), *NGS* next generation sequencing, *PCR* polymerase chain reaction. A detailed protocol for the isolation of plant DNA is provided in Chapter 5.

DNA fingerprinting of plants has become an invaluable tool in forensic (see Chapter 4), scientific, and industrial laboratories all over the world since it was first described in 1988 (for a brief history of DNA fingerprinting see Chapter 6 in ref. 5). Its widespread adoption has been facilitated by the introduction of the polymerase chain reaction (PCR) and heat stable DNA polymerase some 20 years ago (6). PCR has become part of virtually every variation of the plethora of approaches used for DNA fingerprinting today (5). As the following chapters in this volume show, DNA sequencing is increasingly used either in combination with or as a replacement for traditional DNA fingerprinting techniques (Fig. 1). A prime example is the use of short, standardized regions of the genome as taxon barcodes for biological identification of species, which was first proposed by Paul Hebert and colleagues in 2003 (7). Since then, researchers have published more than a thousand barcoding-related papers that have amassed over 14,000 citations (data based on a search of the ISI Web of Knowledge for the topic of "DNA AND barcoding" on 11 June 2011). A Plant Working Group was established at the inaugural meeting of the Consortium for the Barcode of Life (CBOL) and a first proposal for a standardized protocol to barcode all land plants was published in 2007 (8) followed by a proposal of standard DNA barcodes for land plants 2 years later (9). Challenges and detailed methods for barcoding of plants can be found in Chapter 3 and 11 of this volume.

Following the footsteps of automated DNA sequencing (10), "next generation sequencing" (NGS) technology is being developed and employed at a furious pace (11, 12) and is revolutionizing how quickly and cheaply entire genomes can be sequenced (13). Rapid advances in NGS technology are driving down the cost of sequencing and bringing large-scale sequencing projects into the reach of individual investigators (11, 14, 15). NGS technology is beginning to be used to target even the largest nuclear genomes of plants (16–19).

In this chapter, we present an overview of recent publications that demonstrate the use of "NGS" technology for DNA finger-printing and DNA barcoding applications.

2. A Brief Description of Next Generation Sequencing Technology

Currently, the two most commonly used NGS techniques are Pyrosequencing™ used by Roche Diagnostics (formally 454 Sequencing) and Illumina® sequencing (formally Solexa).

2.1. Pyrosequencing™

Pyrosequencing™ is used to determine the sequence of a DNA strand based upon the release of pyrophosphate (PP_i) when a nucleotide is incorporated during DNA synthesis (20, 21). Pyrosequencing™ starts by breaking down the DNA template into many shorter sequences (35–500 bp). Next, the fragments are ligated to adapters that allow the small DNA fragments to bind to the complementary adapter strands bound to beads (one DNA fragment per bead). The bead-bound fragments are then subjected to emulsion PCR for clonal amplification. Beads containing the amplified clonal DNA are then randomly deposited into a micro-fabricated array of wells such that each well contains only one bead. Pyrosequencing™ is then performed on each bead simultaneously. The sequencing reaction takes place by adding sequentially the dNTPs (A, T, C and G) to the polymerase reaction. When a dNTP is successfully incorporated into a nascent DNA strand in a microwell, PP_i is released and converted to ATP using ATP sulfu-rylase and adenosine 5' phosphosulfate. The ATP is subsequently used to drive the conversion of luciferin to oxyluciferin using luciferase, which releases an amount of light proportional to the original amount of PP_i produced (22). An imaging device is used to record the chemiluminescent light signals emanating from each well. The remaining un-incorporated dNTPs are degraded by apyrase to ensure a reaction only occurs upon the addition of new dNTPs. The process is continually repeated until the template

DNA strand has been copied and the intensity profiles of each bead are generated. Currently, this technique generates sequence reads with an average length of about 400 bp (23).

**2.2. Illumina®
Sequencing**

Illumina® sequencing is essentially a highly parallelized adaption of traditional Sanger sequencing (24). For Illumina® sequencing the DNA templates are broken down into short sequences (200–300 bp) to which oligonucleotides containing primer-binding sites are ligated at both ends. Two different adaptors for binding to a flowcell are subsequently attached to the DNA strand using PCR. The DNA is then denatured and run through a flowcell that contains the adapter complements bound to its surface. The single stranded DNA fragments anneal randomly throughout the flowcell to the complementary sites. The surface-bound adapter is then elongated by a polymerase reaction to form a surface-bound clone of the template DNA strand. The DNA is again denatured and washed so only surface-bound DNA strands remain. Bridge amplification is then performed, where the adaptor on the free end of the surface-bound DNA bends down to anneal to an opposite adapter (forming a bridge). The bridged DNA strand is then cloned, thus forming two strands of surface-bound DNA. This process is repeated until approximately 1,000 copies are produced. Bridge amplification is used because the DNA strands are clonally amplified within a very localized area, forming "hot spots" where each unique DNA fragment has been amplified. The sequencing reaction is performed in cycles by adding the dNTPs (all four at the same time) to the flowcell. Each dNTP has a unique, reversibly bound fluorescent label and also serves as a polymerase reaction terminator. Only a single dNTP is therefore incorporated during each polymerase reaction cycle. An imaging device is used to record the colored fluorescent signal generated in the flowcell upon excitation of each DNA "hotspot" by a laser. The reversible terminator on the incorporated nucleotide is then cleaved and the process is repeated until the template strands are copied. The color of each "hotspot" during each nucleotide incorporation cycle is recorded to determine the sequence. Currently, this technique generates sequence reads with an average length of about 40 bp (23).

2.3. Data Processing

Both Pyrosequencing™ and Illumina® sequencing require powerful computers to reconstruct the sequence of the small individual DNA templates at each spot on the array from the imaging data. Each array contains thousands of DNA templates, and consequently NGS typically yields hundreds of millions or even billions of reads during each run.

The relatively short sequence reads obtained by NGS are not sufficient to guide the unequivocal assembly of an entire genome de novo. Therefore, whole genome assembly from NGS data normally requires the availability of an existing reference sequence against which the short NGS reads are aligned. It has recently been reported, however, that a combination of traditional Sanger sequencing, which results in longer sequence reads, and NGS was successfully used to assemble the complete genomes of *Cucumis sativus* (25) and *Vitis vinifera* (26) in the absence of existing reference genomes.

3. NGS of Chloroplast Genomes

Twenty-five years after publication of the first complete sequence of a chloroplast genome by Shinozaki and colleagues (27), several groups have used NGS to sequence the complete chloroplast genomes from multiple plants in parallel. For example, Catherine Nock and colleagues used massively parallel sequencing (MPS) of total DNA and reference-guided assembly using the existing rice genome sequence as a scaffold to assemble the complete chloroplast sequence of wild relatives of increasing evolutionary distance from rice (28). These authors felt that MPS of total DNA for chloroplast genome sequences was simple and cost-effective and suggested that the entire chloroplast genome rather than single genes might in fact serve as the "... the elusive universal single-locus plant barcode for plant species identification." (28). Using multiplexed MPS, Matthew Parks and colleagues reconstructed the infrageneric phylogeny of *Pinus* from 37 nearly complete sequences of chloroplast genomes (120 kb each total length) (29). For multiplex MPS (30, 31), sample-specific indexing or "barcoding" adapters containing sequence tags and a restriction site are ligated to blunt-end repaired DNA samples obtained from a different source (e.g., samples of total DNA obtained from multiple plants). The barcoded samples can then be pooled for MPS. The source of the DNA sequence reads can be determined by the accompanying barcode sequence. Based on their results these authors suspected that "... even the entire chloroplast genome may be insufficient to fully resolve the most rapidly radiating lineages" (29).

Leonie Doorduin and colleagues reported the complete chloroplast genome (150,686 bp) of 17 individuals of the pest species *Jacobaea vulgaris* (32). In this study, chloroplasts from one sample were isolated from 30 g of fresh leaf material for selective extraction of chloroplast DNA. MPS data from the chloroplast DNA was used to assemble a draft chloroplast genome of *J. vulgaris* by aligning the sequence reads to the chloroplast genome reference of *Helianthus annuus*. The authors found only 2% of the reads aligned to the chloroplast reference genome and thus the "purified" chloroplast DNA extract still contained around 98% of non-chloroplast DNA. In an interesting study, J. H. Sønstebø and colleagues used NGS of the P6 loop (a 13–158-bp long variable region with highly conserved flanking sequences) in the chloroplast trnL intron to characterize the composition of past arctic plant communities from a mixture of ancient DNA sequences preserved in permafrost soils (33).

4. NGS in Discovery of Genetic Markers in Plants

Genetic markers in plants include microsatellites (simple sequence repeats, short tandem repeats) (34–38) and single nucleotide polymorphisms (SNPs) (39–43). NGS is rapidly becoming the method of choice for the characterization of genetic markers and the unequivocal identification of multiple alleles of homologous loci in polyploid plants (44).

For example, NGS has been used for the discovery of large numbers of SNPs across the genome of the genus of Japanese rice cultivars (42), *Phaseolus vulgaris* (43), *Populus nigra* (39), and the genus *Vitis* (40). Robert Elshire and coworkers developed a procedure, which they refer to as "genotyping-by-sequencing" approach (45). In their approach, they first extracted high molecular weight DNA from leaves of single maize and barley plants using a standard CTAB protocol and then cut the DNA with methylation-sensitive restriction enzymes that created 2–3-bp overhangs. Methylation-sensitive restriction enzymes don't cut in the heavily methylated repetitive regions of genomes and thus specifically target lower copy regions, which according to the authors significantly simplifies computationally challenging alignment problems in species with high levels of genetic diversity. In the next step, two types of adapters were ligated to the sticky ends of the digested plant DNA. One adapter, referred to as "barcode" adapter terminated with a 4–8-bp barcode on the 3′ end of its top stand and a 3-bp overhang on the 5′ end of its bottom strand that was complementary to the "sticky" end generated by the restriction enzyme used to digest the genomic DNA. The second, or "common," adapter had only the restriction enzyme-compatible sticky end. Following ligation, the restriction fragments with ligated adapters were amplified by PCR using primers that contained at their 3′ end a sequence that was complementary to the ligated barcodes and at their 5′ end a sequence that was complementary to oligonucleotides that coated the flow cell of the NGS machine (Genome Analyzer II; Illumina®, Inc., San Diego, CA) for priming of the subsequent MPS. The MPS data can be aligned to a reference genome (if available, as was in the case of maize) or a reference map is developed around the restriction sites during genotyping of the samples. In the latter case, the consensus reads across the sequence-tagged sites are used as mapping reference. Alternatively, the sequence tags can simply be treated as dominant genetic markers.

5. NGS in Plant Transcriptomics

As Ovidiu Paun and Peter Schönswetter explain in Chapter 7, DNA fingerprinting techniques are not limited to genomic DNA. For example, modifications of the AFLP protocol have been developed for transcriptomic (cDNA-AFLP) and epigenetic studies (methylation-sensitive amplified polymorphism; MSAP). Similarly, NGS can be used to characterize a large number of cellular processes including DNA methylation, histone modification, transcriptome sequencing, alternative splicing identification, small RNA profiling, DNA–protein, and potentially protein–protein interactions (46). For example, NGS has been used for the characterization of the transcriptome of *Cicer arietinum* (47, 48), from 14 different tissues of *Glycine max* (49), *Oryza sativa indica and japonica* (50), *Pinus contorta* (23), *Pisum sativum* (51), and *Scabiosa columbaria* (52). Using NGS, small (short or micro) RNA populations have been characterized at the genome-wide level from uninucleate microspores to tricellular pollen and sporophytic tissues of *Oryza sativa* (53), peanuts (54), and developing tomato fruits (55).

6. Conclusions

The above review of publications demonstrating the use of NGS in plant research show that NGS has the potential to complement or even replace traditional DNA fingerprinting techniques. NGS can also facilitate and complement barcoding of plants. The ability to rapidly and cost-effectively sequence entire chloroplast genomes should be useful in resolving even the toughest barcoding challenges. An important advantage of sequencing-based approaches compared to classical DNA fingerprinting, which is based on electrophoretic band patterns, is that the DNA sequence is independent of the particular method that was used to generate it. Furthermore, DNA sequences can be deposited into public repositories for genetic information such as GenBank, which can be searched easily by other scientists using Web interfaces as well as specialized bioinformatics software (see Chapter 16).

References

1. Herschel WJ (1880) Skin Furrows of the Hand. Nature 23:76
2. Jeffreys AJ, Wilson V, and Thein SL (1985) Individual-specific 'fingerprints' of human DNA. Nature 316(6023):76–79
3. Galton F (1892) Finger prints. Macmillan and Co., New York
4. Hawass Z, Gad YZ, Ismail S, et al (2010) Ancestry and Pathology in King Tutankhamun's Family. JAMA 303(7):638–647
5. Weising K (2005) DNA fingerprinting in plants: principles, methods, and applications. Taylor & Francis Group, Boca Raton
6. Bartlett J and Stirling D (2003) A short history of the polymerase chain reaction. In: Bartlett J & Stirling D (ed) Methods in molecular biology: PCR protocols, 2nd edn. Humana Press Inc., Totowa
7. Hebert PD, Cywinska A, Ball SL, et al (2003) Biological identifications through DNA barcodes. Proc Biol Sci 270(1512):313–321
8. Chase MW, Cowan RS, Hollingsworth PM, et al (2007) A proposal for a standardised protocol to barcode all land plants. Taxon 56(2):295–299
9. Hollingsworth P, Forrest L, Spouge J, et al (2009) A DNA barcode for land plants. Proc Natl Acad Sci USA 106:12794–12797
10. Smith LM, Sanders JZ, Kaiser RJ, et al (1986) Fluorescence detection in automated DNA sequence analysis. Nature 321(6071): 674–679
11. Shendure J and Ji H (2008) Next-generation DNA sequencing. Nat Biotechnol 26(10): 1135–1145
12. Oetting WS (2010) Impact of next generation sequencing: the 2009 Human Genome Variation Society Scientific Meeting. Hum Mutat 31(4):500–503
13. Imelfort M and Edwards D (2009) De novo sequencing of plant genomes using second-generation technologies. Briefings in Bioinformatics 10(6):609–618
14. Zhang J, Chiodini R, Badr A, et al (2011) The impact of next-generation sequencing on genomics. J Genet Genomics 38(3):95–109
15. Varshney RK, Nayak SN, May GD, et al (2009) Next-generation sequencing technologies and their implications for crop genetics and breeding. Trends in Biotechnology 27(9):522–530
16. You FM, Huo N, Deal KR, et al (2011) Annotation-based genome-wide SNP discovery in the large and complex Aegilops tauschii genome using next-generation sequencing without a reference genome sequence. BMC Genomics 12:59
17. Magbanua ZV, Ozkan S, Bartlett BD, et al (2011) Adventures in the enormous: a 1.8 million clone BAC library for the 21.7 Gb genome of loblolly pine. PLoS ONE 6(1):e16214
18. Steuernagel B, Taudien S, Gundlach H, et al (2009) De novo 454 sequencing of barcoded BAC pools for comprehensive gene survey and genome analysis in the complex genome of barley. BMC Genomics 10:547
19. Straub S, Fishbein M, Livshultz T, et al (2011) Building a model: developing genomic resources for common milkweed (Asclepias syriaca) with low coverage genome sequencing. BMC Genomics 12(1):211
20. Nature (2011) Reviews glossary. Nature. http://www.nature.com/nrg/journal/v6/n11/glossary/nrg1709_glossary.html. Accessed 21 June 2011
21. Ronaghi M, Uhlen M, and Nyren P (1998) A sequencing method based on real-time pyrophosphate. Science 281(5375):363–365
22. Ronaghi M, Karamohamed S, Pettersson B, et al (1996) Real-time DNA sequencing using detection of pyrophosphate release. Anal Biochem 242(1):84–89
23. Parchman TL, Geist KS, Grahnen JA, et al (2010) Transcriptome sequencing in an ecologically important tree species: assembly, annotation, and marker discovery. BMC Genomics 11:180
24. Illumina (2010) Technology Spotlight: Illumina® Sequencing.
25. Huang S, Li R, Zhang Z, et al (2009) The genome of the cucumber, Cucumis sativus L. Nat Genet 41(12):1275–1281
26. Velasco R, Zharkikh A, Troggio M, et al (2007) A High Quality Draft Consensus Sequence of the Genome of a Heterozygous Grapevine Variety. PLoS ONE 2(12):e1326
27. Shinozaki K, Ohme M, Tanaka M, et al (1986) The complete nucleotide sequence of the tobacco chloroplast genome: its gene organization and expression. EMBO J 5(9): 2043–2049
28. Nock CJ, Waters DLE, Edwards MA, et al (2011) Chloroplast genome sequences from total DNA for plant identification. Plant Biotechnology Journal 9(3):328–333
29. Parks M, Cronn R, and Liston A (2009) Increasing phylogenetic resolution at low taxonomic levels using massively parallel

sequencing of chloroplast genomes. BMC Biology 7(1):84

30. Meyer M, Stenzel U, and Hofreiter M (2008) Parallel tagged sequencing on the 454 platform. Nat Protoc 3(2):267–278

31. Cronn R, Liston A, Parks M, et al (2008) Multiplex sequencing of plant chloroplast genomes using Solexa sequencing-by-synthesis technology. Nucleic Acids Res 36:e122

32. Doorduin L, Gravendeel B, Lammers Y, et al (2011) The complete chloroplast genome of 17 individuals of pest species Jacobaea vulgaris: SNPs, microsatellites and barcoding markers for population and phylogenetic studies. DNA Res 18(2):93–105

33. Sønstebø JH, Gielly L, Brysting AK, et al (2010) Using next-generation sequencing for molecular reconstruction of past Arctic vegetation and climate. Molecular Ecology Resources 10(6):1009–1018

34. Gardner MG, Fitch AJ, Bertozzi T, et al (2011) Rise of the machines – recommendations for ecologists when using next generation sequencing for microsatellite development. Molecular Ecology Resources. doi:10.1111/j.1755-0998.2011.03037.x

35. Michalczyk IM, Schumacher C, Mengel C, et al (2011) Identification and characterization of 12 microsatellite loci in Cnidium dubium (Apiaceae) using next-generation sequencing. American Journal of Botany 98(5):e127–e129

36. Buehler D, Graf R, Holderegger R, et al (2011) Using the 454 pyrosequencing-based technique in the development of nuclear microsatellite loci in the alpine plant Arabis alpina (Brassicaceae). American Journal of Botany 98(5):e103–e105

37. Delmas CEL, Lhuillier E, Pornon A, et al (2011) Isolation and characterization of microsatellite loci in Rhododendron ferrugineum (Ericaceae) using pyrosequencing technology. American Journal of Botany 98(5):e120–e122

38. Csencsics D, Brodbeck S, and Holderegger R (2010) Cost-Effective, Species-Specific Microsatellite Development for the Endangered Dwarf Bulrush (Typha minima) Using Next-Generation Sequencing Technology. Journal of Heredity 101(6):789–793

39. Marroni F, Pinosio S, Di Centa E, et al (2011) Large scale detection of rare variants via pooled multiplexed next generation sequencing: towards next generation Ecotilling. The Plant Journal. doi:10.1111/j.1365-313X.2011.04627.x

40. Myles S, Chia J-M, Hurwitz B, et al (2010) Rapid Genomic Characterization of the Genus Vitis. PLoS ONE 5(1):e8219

41. Barbazuk WB and Schnable PS (2011) SNP Discovery by Transcriptome Pyrosequencing. cDNA Libraries 729:225–246

42. Arai-Kichise Y, Shiwa Y, Nagasaki H, et al (2011) Discovery of Genome-Wide DNA Polymorphisms in a Landrace Cultivar of Japonica Rice by Whole-Genome Sequencing. Plant and Cell Physiology 52(2):274–282

43. Hyten D, Song Q, Fickus E, et al (2010) High-throughput SNP discovery and assay development in common bean. BMC Genomics 11(1):475

44. Griffin PC, Robin C, and Hoffmann AA (2011) A next-generation sequencing method for overcoming the multiple gene copy problem in polyploid phylogenetics, applied to Poa grasses. BMC Biol 9:19

45. Elshire RJ, Glaubitz JC, Sun Q, et al (2011) A Robust, Simple Genotyping-by-Sequencing (GBS) Approach for High Diversity Species. PLoS ONE 6(5):e19379

46. Lister R, Gregory BD, and Ecker JR (2009) Next is now: new technologies for sequencing of genomes, transcriptomes, and beyond. Current Opinion in Plant Biology 12(2):107–118

47. Molina C, Zaman-Allah M, Khan F, et al (2011) The salt-responsive transcriptome of chickpea roots and nodules via deepSuperSAGE. BMC Plant Biology 11(1):31

48. Hiremath PJ, Farmer A, Cannon SB, et al (2011) Large-scale transcriptome analysis in chickpea (Cicer arietinum L.), an orphan legume crop of the semi-arid tropics of Asia and Africa. Plant Biotechnology Journal:10.1111/j.1467-7652.2011.00625.x

49. Severin A, Woody J, Bolon Y-T, et al (2010) RNA-Seq Atlas of Glycine max: A guide to the soybean transcriptome. BMC Plant Biology 10(1):160

50. Lu T, Lu G, Fan D, et al (2010) Function annotation of the rice transcriptome at single-nucleotide resolution by RNA-seq. Genome Research 20(9):1238–1249

51. Franssen S, Shrestha R, Brautigam A, et al (2011) Comprehensive transcriptome analysis of the highly complex Pisum sativum genome using next generation sequencing. BMC Genomics 12(1):227

52. Angeloni F, Wagemaker CAM, Jetten MSM, et al (2011) De novo transcriptome characterization and development of genomic tools for Scabiosa columbaria L. using next-generation sequencing techniques. Molecular Ecology Resources 11(4):662–674

53. Wei LQ, Yan LF, and Wang T (2011) Deep sequencing on genome-wide scale reveals the unique composition and expression patterns of microRNAs in developing pollen of Oryza sativa. Genome Biol 12(6):R53

54. Zhao C-Z, Xia H, Frazier T, et al (2010) Deep sequencing identifies novel and conserved microRNAs in peanuts (Arachis hypogaea L.). BMC Plant Biology 10(1):3

55. Mohorianu I, Schwach F, Jing R, et al (2011) Profiling of short RNAs during fleshy fruit development reveals stage-specific sRNAome expression patterns. The Plant Journal. doi:10.1111/j.1365-313X.2011.04586.x

Chapter 3

Challenges in the DNA Barcoding of Plant Material

Robyn S. Cowan and Michael F. Fay

Abstract

DNA barcoding, using a short gene sequence from a standardized region of the genome, is a species identification tool which would not only aid species discovery but would also have applications ranging from large-scale biodiversity surveys through to identification of a single fragment of material in forensic contexts. To fulfill this vision a universal, relatively cheap, scalable system needs to be in place. The mitochondrial locus being used for many animal groups and algae is not suitable for use in land plants, and an appropriate alternative is needed.

Progress has been made in the selection of two alternative regions for plant DNA barcoding. There are however many challenges in finding a solution that fulfills all the requirements of a successful, universally applicable barcode, and in the short term a pragmatic solution that achieves as much as possible and has payoffs in most areas has been chosen. Research continues in areas ranging from the technicalities of sequencing the regions to data analysis and the potential improvements that may result from the developing technology and data analysis systems.

The ultimate success of DNA barcoding as a plant identification tool for all occasions depends on the building of a reference database and it fulfilling the requirements of potential users such that they are able to achieve valid results through its use, that would be more time consuming and costly, and less reliable using other techniques.

Key words: DNA barcoding, Land plants, Plastid, CBOL, Data analysis, *rbcL*, *matK*

1. Introduction

In this chapter on DNA barcoding in land plants, we discuss its development and the many challenges involved in implementing a successful barcoding system. Some of these challenges, such as data analysis and referencing, are common to DNA barcoding in all organisms, whereas others are specific to land plants, particularly the choice of the most appropriate DNA region or regions for use as the barcode.

"DNA barcoding" was proposed in 2003 as the use of a short gene sequence from a standardized region of the genome as a tool for

Nikolaus J. Sucher et al. (eds.), *Plant DNA Fingerprinting and Barcoding: Methods and Protocols*, Methods in Molecular Biology, vol. 862, DOI 10.1007/978-1-61779-609-8_3, © Springer Science+Business Media, LLC 2012

species identification (1, 2), with the mitochondrial gene cytochrome c oxidase, subunit I, (*cox1* or CO1) being used as a barcode in Lepidoptera. This was soon followed by studies that claimed a high success rate in species identification using *cox1* in other groups such as birds (3), fish (4), and mosquitoes (5). There has been a steady flow of publications on DNA barcoding in different groups ever since.

In 2004, the Consortium for the Barcode of Life (CBOL) was founded, with funding from the Alfred P. Sloan Foundation, as an international initiative to promote and develop DNA barcoding (http://www.barcoding.si.edu, http://www.barcodeoflife.org). At the inaugural meeting of CBOL, a Plant Working Group (PWG) was formed to explore the particular challenges of DNA barcoding in plants, more specifically land plants, for which *cox1* holds little potential as a suitable DNA barcode region. For some reason the journey of plants (algae) from watery habitats to the land coincided with a change in the way that mitochondrial genomes behave and evolve. In animals there is a high base-substitution rate, but the gene content and order are highly conserved; whereas in plants, with a few exceptions in specific taxa (6–8), base substitution rates are much lower, and there are frequent genome rearrangements, transfers of genes between different genomes (plastid, mitochondrial, and nuclear) and across species (9, 10). However, particularly since the development of the polymerase chain reaction (PCR) and Sanger sequencing, there has been a wealth of molecular phylogenetic studies of various plant groups using plastid and nuclear genome sequence data, with a study across the angiosperms using the large subunit of the ribulose-bisphosphate carboxylase gene (*rbcL*) of plastid DNA (11) being groundbreaking in scope. In addition to elucidation of evolutionary relationships, DNA data (including sequences and DNA fingerprinting techniques such as plastid and nuclear microsatellites and AFLP) have also been used to study population dynamics, e.g., (12), species delimitation, e.g., (13), hybridization, e.g., (14), and in an ad hoc way to identify samples when morphological characters are insufficient, e.g., the identification of an unknown plant (15) or pollen on a bee (16). Although these examples exhibit the great utility of DNA sequence data, a major drawback relevant to DNA barcoding has been that the techniques and data used were not standardized or in many cases amenable to standardization, a primary requirement for DNA barcoding.

DNA barcoding can be used for species discovery and delimitation, large scale floristic inventories and monitoring changes over time, identification from different plant parts (e.g., seeds or roots) or fragments of material where morphological characters are insufficient (forensics, biocontrol, CITES), authentication of food and herbal medicines, elucidation of historical patterns of distribution, and study of plant–animal interactions (e.g., which species is living on which plant species, or what the diet of an animal is by identifying its stomach contents). However, to realize this potential, particularly in a readily accessible and cost-effective manner, a number of characteristics are necessary or at least highly desirable in a DNA barcode, and the first major challenge for plants was the identification of a DNA region or regions that showed these characteristics.

2. Desired Characteristics of a DNA Barcode

The most important characteristics if the system is to be universally applicable are probably to be able to achieve universal amplification, i.e., across all taxa, using standardized primers, and being technically simple to sequence and short enough to sequence in one reaction, which with present-day capillary sequencing is ~750 bp (with advances in sequencing technology modifications of this requirement may be possible; see below). In addition, the barcode needs to exhibit sufficient variability for species-level identification, ideally with high interspecific and low intraspecific sequence divergence. Together, these allow easy referencing and species delimitation. To facilitate analysis, the barcode should be easily alignable, one DNA sequence against another, and should contain few insertions/deletions, as these complicate the comparison and can be difficult to interpret. Also, for many potential applications the barcode needs to be readily recoverable from herbarium samples and other degraded samples (17).

3. Barcode Selection

With the limitations of the mitochondrial genome as the source of a suitable DNA barcode discussed earlier, the nuclear and plastid genomes were both considered. However, use of the biparentally inherited nuclear genome requires the ability to "disentangle" different alleles of the region being sequenced. It is also more difficult to recover nuclear DNA suitable for sequencing from herbarium samples and other degraded samples than is plastid DNA. This is possibly, at least in part, because there are many more copies per cell of the plastid genome and because of differences in the protection afforded by the nuclear and plastid membranes. The nuclear internal transcribed spacer region (ITS), a tandem repeat that is generally more successfully retrieved from degraded DNA than, for example, single- or low-copy nuclear genes and has commonly been used in species-level plant phylogenetic studies, was identified as a potential barcode region by Kress et al. (18). This region generally shows high levels of interspecific sequence variability, but there are problems with paralogues and the maintenance of multiple uncorrected copies in many plant groups, making it impossible to sequence the region directly without cloning. The presence of paralogues also presents a problem of ensuring a "like for like" comparison across samples and taxa.

Considerable effort was therefore concentrated on examining the plastid genome, both coding and non-coding regions, for a potential DNA barcode. Ford et al. (19) screened 41 of the 81 coding regions of the *Nicotiana* plastid genome using criteria of (1) being routinely amplifiable, (2) being easily sequenced using single-pass sequencing, (3) having the highest possible sequence variability but not being susceptible to amplification of any other than the target plastid region, e.g., the inadvertant amplification of nuclear pseudogenes, and (4) being easily annotated for quality checking. From this, an eventual shortlist of five regions was tested further and a proposal was made that a three-locus barcode would be most appropriate (17). Meanwhile, Kress et al. (20), in addition to their evaluation of ITS, also investigated two plastid regions, the gene *rbcL* and the non-coding spacer region *trnH-psbA*. Based on whole plastid genome sequencing of *Panax* and 17 other vascular plants, Kim and Lee (21) proposed two non-coding spacer regions, *atpF-atpH* and *trnH-psbA*. Another plastid region that has been proposed is the *trnL* intron (22). However, while this locus has been shown to work as an identification aid in some situations, e.g., ancient wood (23) or some bryophytes (24), it has an overall relatively slow rate of molecular evolution and is surprisingly conserved for a non-coding region (25). A final evaluation of the resulting seven leading candidate regions under the auspices of the PWG led to the proposal that partial (single-read) regions of *rbcL* and maturase K (*matK*) be used as a two-locus plant DNA barcode. Finding a barcode that completely fulfilled the desired characteristic criteria was not entirely achievable and this decision was based on a complex trade-off of these, especially universality, sequence quality, power to discriminate among species, and cost (26). Several combinations of two loci gave 70–75% species discrimination in the sample set used for the evaluation, and adding further loci did not significantly increase this, although much higher levels will be found in studies within a limited geographical area (22, 27). One of the loci, *rbcL*, was easily retrievable across the land plants, but the other, *matK*, with the highest species discrimination among the coding regions tested, was not so easily retrievable using universal primers, particularly in the non-angiosperm samples. On the other hand, the non-coding regions exhibited poor sequence quality, mainly caused by microsatellite regions and consequent stutter (28). Overall it was considered more likely that it would be possible to improve the recoverability of *matK* through the improvement of primer universality than to overcome the sequence quality issues of the tested non-coding regions.

CBOL accepted the proposal from the PWG for a two-locus barcode (*rbcL* plus *matK*), http://www.barcoding.si.edu/plant_working_group.html. The PWG recognised that, given the limitations of the two-locus barcode, other non-coding regions were likely to be useful to enhance the discriminatory power of the barcode in particular taxa. The PWG at this time also posted information on the most successful methods and primers available at the time for using these regions as barcodes (http://www.barcoding.si.edu/PDF/Informationonbarcodeloci.pdf), although work continues on improving these.

4. Problems and Challenges

There are three main areas in which the chosen barcode regions present problems: taxonomic congruence, the implications of using plastid regions and the degree to which they fall short of fulfilling the desired characteristics discussed above.

4.1. Taxonomic Congruence

The concept of DNA barcoding relies on a high degree of congruence between the barcodes and the species taxonomy of the group in question. Ideally there will be more interspecific than intraspecific variation, but perhaps more importantly there should be no overlap of absolute barcode sequence between species, i.e., that there is a perfect one-to-one match between barcodes and taxa. To date, these criteria are fulfilled in a lower percentage of plant than animal studies. Fazekas et al. (29) discussed whether this may be due not only to lower variability in the potential plant DNA barcode regions but also because the boundaries between plant genomes are less well defined, citing such reasons as hybridization and polyploidy, incomplete sorting of ancestral polymorphisms, and imperfect taxonomies and species definitions.

Many groups of plants include species with little or no plastid sequence divergence from each other, e.g., species-rich genera in which rates of species identification using a variety of putative barcodes rarely exceed 70% (30) and additional information may be required. In some cases this may be as simple as the geographical origin of the sample. For example, they were able to increase this percentage to >98% with a three-locus barcode in the case of woody plants from Barro Colorado island (Panama) for which a complete species list is available. Similarly (although not set in a barcoding context), Richardson et al. (31, 32) obtained more or less identical DNA sequences for three species of *Phylica* (Rhamnaceae), which are products of a relatively recent radiation. However, as each of these is only known from one geographical region (South Africa, the

Tristan da Cunha archipelago plus Amsterdam Island or St Helena), knowledge of the origin of the sample allowed 100% species-level identification.

4.2. Implications of Using Plastid Regions

As the plant DNA barcode is made up of two plastid regions, hybridization and polyploidy have a predictable effect on species identification which is potentially greater in groups with high levels of apomixis.

4.3. Hybridization

Hybrids inherit the plastid genome of one parent only (in the vast majority of angiosperms, this is from the maternal parent, whereas in conifers, for example, it is from the paternal parent), and once hybridization has taken place, backcrosses to the other parent can result in "plastid capture," i.e., the introgression of plastids from one species into another e.g. (33). After several generations of backcrossing, the resulting progeny may possess the more or less "pure" nuclear genome of one species, but the plastid genome of the second species. Under these circumstances, use of the standard plant barcoding loci (and any other plastid loci) will give the "wrong" answer. Numerous examples of this type of situation can be found in the scientific press.

One good example occurs in the genus *Orchis* (Orchidaceae). *Orchis militaris, Orchis purpurea,* and *Orchis simia* are three closely related species of European terrestrial orchids, and when two or more of these occur together, they can hybridize and produce progeny which are clearly intermediate in morphology. Molecular data, however, have demonstrated that some populations that have been believed to be pure examples of one species on the basis of morphology and nuclear DNA studies have the plastid genome of one of the other species. For example, some populations of *O. purpurea* and one population of *O. simia* in England have the plastid genome of *O. militaris*, and barcoding of these populations would identify them as *O. militaris* (34). In these species, data collected so far indicate that this hybridization (and the resulting plastid capture) is more or less unidirectional, with *O. militaris* being the plastid donor in the vast majority of cases.

Other examples of introgression (out of many in the literature) between species leading to plastid capture have been reported for the *Heuchera* group (Saxifragaceae) (35) and native and introduced *Juglans* spp. (Juglandaceae) in North America (36). In studies examining the possibility of geneflow from genetically modified oilseed rape (*Brassica napus,* a tetraploid member of Brassicaceae) to related species, hybridization and plastid capture were detected between *B. napus* and weedy *Brassica rapa* (a diploid), despite the difference in ploidy, resulting in *B. rapa*-like plants having *B. napus* plastids (37, 38).

4.4. Polyploidy

Polyploidy is a common phenomenon in angiosperms with >70% of species having been suggested to have one or more polyploid events in their history (39), and this can also lead to problems with barcoding based on the plastid genome. Recently formed autopolyploids (which can be called by the same name as the diploid progenitor or be given a different name, depending on taxonomic preference in the group in question) will have the same plastid genome as the parental diploid. With increasing time, due to reproductive isolation that occurs in most (or many) cases involving different ploidy, differences in the plastid genome may accumulate, but at least in the early generations following polyploidization, barcoding will fail to discriminate between the diploid and the polyploid lineages. Recently formed allopolyploids will have the plastid genome of one of the progenitors only, and, for this reason, barcoding alone will fail to distinguish between the plastid donor parent and the polyploid. A combination of morphological study and barcoding may provide a solution for allopolyploids when sufficiently intact specimens are available for examination (in some cases), but autopolyploids and parental diploids are often indistinguishable on the basis of gross morphology and have been treated as cytotypes of the same species, e.g., (40). In groups where apomictic individuals are common and polyploid (e.g., in *Rubus*, *Sorbus*, *Taraxacum*, and *Limonium*), the plastid genome in the apomicts is likely to be "frozen" and therefore even old polyploids may not be distinguishable from one of the progenitors. In *Sorbus*, a study of >450 individuals of three apomictic complexes and their diploid progenitors showed that for all individuals tested in the *Sorbus anglica* and *Sorbus latifolia* apomictic complexes (derived from *Sorbus aria s.l.* and *Sorbus aucuparia* and *S. aria s.l.* and *Sorbus torminalis*, respectively) *S. aria s.l.* was the pollen donor (i.e., male parent) and as a result had the plastid type of the other parent (either *S. aucuparia* or *S. torminalis*) (41). In cases like this, use of plastid DNA barcodes will inevitably fail to distinguish between the maternal diploids and their polyploid derivatives.

The fact that DNA barcoding will fail in situations like this, has led to the suggestion that a "traffic light" system (42, 43) similar to systems used with commercial barcodes (e.g. http://www.isisintegration.co.uk/isis_checkriteIII.html) could be used. A green light would indicate a certain match, e.g., a DNA barcode which matched perfectly that of a species known to have a unique barcode. Members of isolated lineages or of lineages that have been extensively studied with barcoding at the species level would be likely to get a green light. An amber light would indicate "proceed with caution" (e.g., the barcode matches a small number of closely related species or a group in which hybridization is known to occur, e.g., *Orchis* as discussed above) and a red light would indicate that the barcoding approach does not work with this particular taxon. This could be due to various reasons including no match in the existing database, poor quality of the submitted sequence, or multiple matches (as would be the case with *Sorbus* mentioned above). Inevitably, the quality of such a traffic light system will depend on the quality and completeness of the barcode database and an understanding of genetic patterns (e.g., prevalence of hybridization and polyploidy) in the group in question.

5. Reference Database/Analysis

In order to be able to identify an unknown sample from its DNA barcode, it is necessary to have reference barcodes available from fully identified samples for comparison. These need to include the species in question, ideally with multiple samples that cover all intraspecific variability and all closely related species that may have overlapping or similar barcodes. Much of the international DNA barcoding effort at present involves "building" such a reference database and there are two main conduits for this. Firstly, GenBank (44) and its counterparts, the DNA DataBank of Japan (DDBJ), and the European Molecular Biology Laboratory (EMBL) form the primary repository for the reference barcodes. The National Center for Biotechnology (NCBI), a U.S. government-funded national resource for molecular biology information responsible for the GenBank DNA sequence database, has developed a barcode bulk submission tool (to date only for *cox1*) which also allows for the submission of further data, including but not limited to voucher information, verifier, collection location, and DNA barcode sequence trace files. If sufficient data requirements are fulfilled, the keyword "barcode" is attached to the record.

Secondly, and complimentary to this facility, the Barcode of Life Data Systems (BOLD) "is an online workbench that aids collection, management, analysis, and use of DNA barcodes" (45). This is open for use as a platform for all researchers in DNA barcoding to upload and manage the data associated with their projects. The researchers have control over their own data including its public release. The plant DNA barcode regions have been incorporated into BOLD and automatic download from BOLD to GenBank will become available.

If DNA barcoding is to be of widespread use, it is crucial that the reference database contains high-quality sequences that are correctly identified. It is easy to imagine the possible consequences of misidentification in forensic cases such as a possible miscarriage of justice, the introduction of a lethal invasive species that was identified as another benign species, or the financial collapse of an enterprise that incorrectly had products seized as breaking CITES regulations, and incorrect identifications on reference barcode sequences is one way this could occur. It is clear that there are a considerable number of incorrectly labeled sequences in GenBank, and this can happen for a number of reasons, including contamination of the original DNA sample, e.g., using ITS primers which are universal across plants and fungi to sequence a plant DNA sample that is contaminated with endogenous fungal material, incorrect identification of the original material from which the DNA was extracted, or poor quality DNA sequences which are then misinterpreted. The requirements for the "barcode" keyword aim to reduce the risk of these problems occurring and to provide a better data trail for retrospective checking and facilitate a process by which the "barcode" label can be removed if it is shown that the barcode sequence is incorrect in any way.

Assuming the existence of a reference database of high-quality, correctly identified sequences, there is still the requirement for a query system that can match a DNA barcode from an unknown sample to the correct species using these reference barcodes. The most commonly used query methods in barcoding fall broadly into two camps: similarity methods such as BLAST, megaBLAST, and FASTA, all of which return a result relatively quickly, and clustering methods such as parsimony and neighbor joining, which are much slower and probably not realistic for a universal plant query system at present. These methods are also prone to error and inconsistencies especially when dealing with some of the situations that occur in barcoding.

An incomplete reference dataset will reduce the probability of a match but might also assign an unknown to the wrong species. If two or more species share the same barcode and they all have reference barcodes, it is not possible to identify the unknown as a species, but it can be allocated to a small group of species. However, if one or more of these overlapping species is/are missing from the barcode database, a complete misidentification could occur. Errors may also occur where intraspecific variability is as high or higher than interspecific variability, and a given query is more similar to the barcodes of another species than the one to which it correctly belongs. Alternative query methods are being investigated, and an overview of the limitations of the present systems and some of the alternative methods can be found in (46), although none has been developed as a publicly available system that can be used for routine queries on large datasets as yet.

References

1. Hebert PDN et al (2003) Biological identifications through DNA barcodes. Philos Trans Royal Soc B 270:313–321.

2. Hebert PDN, Ratnasingham S and De Waard JR (2003) Barcoding animal life: cytochrome c oxidase subunit 1 divergences among closely related species. Philos Trans Royal Soc B 270:S96–S99.

3. Hebert PDN et al (2004) Identification of birds through DNA barcodes. PLoS Biol 2:e312. doi:10.1371/journal.pbio.0020312.

4. Ward RD, et al (2005) A start to DNA barcoding Australia's fish species. Philos Trans Royal Soc, Ser B 360:1847–1857.

5. Cywinska A, Hunter FF, Hebert PDN (2006) Identifying Canadian mosquito species through

DNA barcodes. Med Vet Entomol 20: 413–424

6. Bakker FT, et al (2000. Mitochondrial and chloroplast DNA-based phylogeny of *Pelargonium* (Geraniaceae). Amer J Bot 87:727–734.

7. Cho Y, et al (2004) Mitochondrial substitution rates are extraordinarily elevated and variable in a genus of flowering plants. Proc Natl Acad Sci U.S.A. 101:17741–17746.

8. Parkinson P L, et al (2005) Multiple major increases and decreases in mitochondrial substitution rates in the plant family Geraniaceae. BMC Evol Biol 5:73.

9. Palmer J D, et al (2000) Dynamic evolution of plant mitochondrial genomes: mobile genes and introns and highly variable mutation rates. Proc Natl Acad Sci U.S.A. 97:6960–6966.

10. Mower J P, et al (2004) Plant genetics: gene transfer from parasitic to host plants. Nature 432:165–166.

11. Chase MW, et al (1993) Phylogenetics of seed plants: an analysis of nucleotide sequences from the plastid gene *rbcL*. Ann Mo Bot Gard 80: 528–548+550–580.

12. Fay MF, Krauss SL (2003) Orchid conservation genetics. in the molecular age. In: Dixon KW, et al (eds.), Orchid conservation. Natural History Publications, Kota Kinabalu, Sabah.

13. Richardson JE, et al (2003) Species delimitation and the origin of populations in island representatives of *Phylica* (Rhamnaceae). Evolution 57:816–827.

14. Clarkson JJ, et al (2004) Phylogenetic relationships in *Nicotiana* based on multiple plastid loci. Molec Phylog Evol 33:75–90.

15. Bradford JC, Barnes RW (2001) Phylogenetics and classification of Cunoniaceae (Oxalidales) using chloroplast DNA sequences and morphology. Syst Bot 26:354–385.

16. Widmer A, et al (2000) Molecular analysis of orchid pollinaria and pollinaria-remains found on insects. Mol Ecol 9:1911–1914.

17. Chase MW, et al (2007) A proposal for a standardised protocol to barcode all land plants. Taxon 56:295–299.

18. Kress JW, et al (2005) Use of DNA barcodes to identify flowering plants. Proc Natl Acad Sci 102:8369–8374.

19. Ford CS, et al (2009) Selection of candidate coding DNA barcoding regions for use on land plants. Bot J Linn 159:1–11.

20. Kress WJ, Erickson DL (2007) A two-locus global DNA barcode for land plants: the coding *rbcL* gene complements the non-coding

trnH-psbA spacer region. PLoS ONE 2: e508. doi:10.1371/journal.pone.0000508.

21. Kim KJ, Lee H-L (2004) Complete chloroplast genome sequences from Korean Ginseng (*Panax schinseng* Nees) and comparative analysis of sequence evolution among 17 vascular plants. DNA Research 11:247–261.

22. Taberlet P, et al (2007) Power and limitations of the chloroplast *trn*L (UAA) intron for plant DNA barcoding Nucl Acids Res 35:e14 doi:10.1093/nar/gkl938.

23. Liepelt S, et al (2006) Authenticated DNA from ancient wood remains. Ann Bot 98:1107–1111

24. Rowntree JK et al (2010) Which moss is which? Identification of the threatened moss *Orthodontium gracile* using molecular and morphological techniques. Conser Genet 11:1033–1042.

25. Shaw J, et al (2005) The tortoise and the hare II: Comparison of the relative utility of 21 non-coding chloroplast DNA sequences for phylogenetic analysis. Amer J Bot 92:142–166.

26. CBOL Plant Working Group (2009) A DNA barcode for land plants. Proc Natl Acad Sci U.S.A. 106:12794–12797.

27. Janzen DH (2005) in Plant conservation: A natural history approach. Krupnick G, Kress WJ (eds) University of Chicago Press, Chicago.

28. Devey DS, Chase MW, Clarkson JJ (2009) A stuttering start to plant DNA barcoding: microsatellites present a previously overlooked problem in non-coding plastid regions. Taxon 58:7–15.

29. Fazekas AJ, et al (2009) Are plant species inherently harder to discriminate than animal species using DNA barcoding markers?. Mol Ecol Resour 9:130–139.

30. Kress WJ, et al (2009) Plant DNA barcodes and a community phylogeny of a tropical forest dynamics plot in Panama. Proc Natl Acad Sci U.S.A. 106:18621–18626.

31. Richardson JE, et al (2001) Rapid and recent origin of species richness in the Cape Flora of South Africa. Nature 412:181–183.

32. Richardson JE, et al (2001) Phylogenetic analysis of *Phylica* L. with an emphasis on island species: evidence from plastid *trn*L-F DNA and nuclear internal transcribed spacer (ribosomal DNA) sequences. Taxon 50:405–427.

33. Tsitrone A, Kirkpatrick M, Levin DA (2003) A model for chloroplast capture. Evolution 57:1776–1782.

34. Fay MF, et al (2007) How does hybridization influence the decision making process in conservation? The genus *Orchis* (Orchidaceae) as a case history. Lankesteriana 7:135–137.

35. Soltis DE, Kuzoff RK (1995) Discordance between nuclear and chloroplast phylogenies in the *Heuchera* group (Saxifragaceae). Evolution 49:727–742.

36. Hoban SM et al (2009) Geographically extensive hybridization between the forest trees American butternut and Japanese walnut. Biol Letters 5:324–327.

37. Hansen LB, Siegismund HR and Jørgensen RB (2003) Progressive introgression between *Brassica napus* (oilseed rape) and *B. rapa*. Heredity 91:276–283.

38. Haider N, Allainguillaume J and Wilkinson MJ (2009) Spontaneous capture of oilseed rape (*Brassica napus*) chloroplasts by wild *B. rapa*: implications for the use of chloroplast transformation for biocontainment. Curr Genet 55:139–150.

39. Meyers LA, Levin DA (2006). On the abundance of polyploids in flowering plants. Evolution 60:1198–1206.

40. Soltis DE, et al (2007) Autopolyploidy in angiosperms: have we grossly underestimated the number of species? Taxon 56:13–30.

41. Chester M, et al (2007) Parentage of endemic *Sorbus* L. (Rosaceae) species in the British Isles: evidence from plastid DNA. Bot J Linn Soc 154:291–304.

42. Chase MW, et al (2005) Land plants and DNA barcodes: short-term and long-term goals. Philos Trans Royal B 360:1889–1895.

43. Cameron, K. University of Wisconsin, pers. comm.

44. Benson DA, et al (2010) GenBank. Nucleic Acids Res 38:D46–D51.

45. Ratnasingham S, Hebert PDN (2007) BOLD : The Barcode of Life Data System (www.barcodinglife.org). Mol Ecol Notes 7: 355–364.

46. Little DP, Stevenson DW (2007) A comparison of algorithms for the identification of specimens using DNA barcodes: examples from gymnosperms. Cladistics 23:1–21.

Chapter 4

Plant Genetics for Forensic Applications

David N. Zaya and Mary V. Ashley

Abstract

An emerging application for plant DNA fingerprinting and barcoding involves forensic investigations. Examples of DNA analysis of botanical evidence include crime scene analysis, identifying the source of commercial plant products, and investigation of trade in illicit drugs. Here, we review real and potential applications of DNA-based forensic botany and provide a protocol for microsatellite genotyping of leaf material, a protocol that could be used to link a suspect to a victim or to a crime scene.

Key words: Forensic botany, Microsatellites, Crime scene investigation, Plant product identification, DNA barcoding

1. Introduction

Plant DNA fingerprinting is an under-utilized tool for forensic applications, but the potential for DNA profiling of botanical specimens to contribute forensic evidence is an important, emerging field. Depending upon the crime, assignment of botanical material to a particular species may aid an investigation and provide important evidence. Such species-assignment requires a barcoding approach and the establishment of a database of species-specific sequences at one or more genetic loci—often on the chloroplast genome. Alternatively, if the source species is identified, it may be possible to match botanical material to an individual plant, clone, or population, through DNA fingerprinting. This could be accomplished by using multilocus genotypes at highly variable loci such as DNA microsatellites, also known as simple sequence repeats (SSRs), short tandem repeats (STRs), or variable number tandem repeats (VNTRs). We briefly review both of these general approaches, then provide examples of a wide range-of forensic applications utilizing these approaches.

Nikolaus J. Sucher et al. (eds.), *Plant DNA Fingerprinting and Barcoding: Methods and Protocols*, Methods in Molecular Biology, vol. 862, DOI 10.1007/978-1-61779-609-8_4, © Springer Science+Business Media, LLC 2012

Botanical evidence such as leaves or seeds can often be assigned to species using morphology and plant anatomy. However, identification based on physical characteristics alone may be ambiguous or uncertain, can only be done by a skilled botanist trained in taxonomy, and the evidence material must be largely intact. Trace evidence, such as leaf fragments, seeds, or pollen, is nearly impossible to assign to species based on morphology, and may not even be recognized as possible evidence by crime scene investigators. However, DNA can be extracted from even tiny fragments of plant material, and sequence data obtained from this material can potentially be used to assign the evidence unambiguously to species. DNA barcoding involves the amplification and sequencing of relatively short, standardized genetic loci, together with the establishment of a database of sequences at these loci (DNA barcodes) that allow rapid and accurate identification of species. The DNA region chosen for barcoding should have enough variability to distinguish species in nature, but also have conserved region for development of universal primers. It also should show little intraspecific sequence variation. For animals, the mitochondrial cytochrome c oxidase I gene has become established as a universal barcode. Unfortunately, no single region in the plant genome has emerged as a likely barcode candidate (1, 2). The mitochondrial genome of plants is extremely conserved and exhibits inadequate variation for barcoding. Ferri et al. (3) tested two regions of the chloroplast genome (psbA-trnH and trnL-trn F) and showed that these loci allowed the species resolution of 60% of the samples tested from a wide variety of plant species. Focusing on grasses, Ward et al. (4) have developed a series of taxon-specific PCR assays involving six mitochondrial and chloroplast loci for a molecular identification system, but conclude that no single locus can be used to identify all grass species. While challenges remain in developing a barcoding approach that can be widely used in forensic botany, the accumulation of sequence databases for plant genomes can already aid in species identification of botanical evidence. This is especially true if investigators are trying to distinguish DNA from a limited number of species.

DNA fingerprinting using multilocus genotypes can provide much higher resolution than barcoding. Botanical evidence can be traced to a specific source plant, and thus to an exact location. The approach has great potential, but as with barcoding, technical obstacles remain. Microsatellite primers are not universal, and may only work with a relatively small group of closely related species, such as congeners. For many plant species, DNA fingerprinting will first require the screening of genomic libraries for STRs, followed by optimization and screening of primer pairs for reliable amplification and adequate variability. Microsatellite genotyping and individual identification may also be complicated by the fact that many plant species are polyploid, and in polyploids the number of alleles expected at each locus is not straightforward. Polyploidy will also complicate inferences of important statistics for forensics, such as the probability of identity (PPI) or the polymorphic information content (PIC) for microsatellite loci. Even with these obstacles, however, the potential of DNA fingerprinting in forensics has already been clearly demonstrated in several studies.

To provide an overview of plant genetics for forensic applications, we describe examples from three quite distinct applications: (1) crime scene investigations, (2) identification of plant products, and (3) identification of the source of illegal narcotics.

1.1. Crime Scene Investigations

Botanical evidence found at crime scenes or in association with victims or suspects has the potential to provide valuable evidence for criminal investigations. Plant DNA has the potential to establish linkages between the victim, the suspect, or the scene of a crime. For example, plant material taken from a victim might be used to identify where a crime was committed, or DNA isolated from seeds or leaves from a suspect's clothing or automobile might place the suspect at a crime scene. One of the first examples of a DNA-based forensic botany investigation involved a homicide, where seed pods of the Palo Verde tree (*Cerdicium* sp.) were recovered from the vehicle of a suspect. Randomly amplified polymorphic DNA (RAPD) markers provided a genetic profile to link the seed pods to a single tree at the murder site (5, 6). In another homicide case, three species of bryophytes (mosses) were found on the suspect and identified to species. DNA fingerprinting of the bryophyte material demonstrated that it likely originated from the crime scene (7). Another homicide was investigated using leaves of sand live oak, *Quercus geminata*, collected from three trees at site where the victim was found, and those found in the trunk of a suspect's car (8). While the leaves from the suspect's car did not match the trees at the crime scene, this study: (1) demonstrated the successful genotyping of evidence leaves that were several years old and (2) showed that using only four microsatellite loci, the probability that two trees would have the same genetic profile were only 1 in 500,000. To make such statistical inference, an adequate sample of plants of the same species in the same region as the crime scene must be genotyped to provide population statistics on genetic variability and allele frequencies. A protocol for the microsatellite genotyping method used in the *Q. geminata* study is provided in this chapter.

Quite a different application involves analysis of plant DNA from the gastric contents of a deceased person, to aid in investigating the cause of death (9–11). For example, sequences of chloroplast genes identified dandelion DNA in a decedent's stomach, suggesting that dandelions were mixed with sleeping drugs to avoid detection (11).

1.2. Plant Product Forensics

The use of DNA fingerprinting has been applied to identification of the species, variety, or geographical source of commercially available plant products. Much of the work in this field has concentrated on the label accuracy or adulteration of herbal medications, the source species or region of timber products, and the presence of allergens in food. The common theme is the need for molecular markers, because after processing and manufacturing the product is impossible to identify by physical appearance. The DNA techniques used vary greatly, depending largely on the goal of the investigation and the resolution required. It should be noted that plant products may have highly degraded DNA—due to processing, age, or storage conditions (12). Thus, molecular markers that rely on shorter DNA sequences may be best suited for analysis of highly processed plant products (13).

When testing plant products, researchers usually search for the presence of a single species or a predetermined list of species. This is often the case when working with herbal medication, for which there are few checks on sale and labeling. Published reports generally concentrate on developing molecular tools that can distinguish the target ingredient(s) from ineffective substitutes, or detect adulterants that are ineffective or harmful. DNA tools have been extensively applied to identification of herbal medication, especially those plants used in traditional Chinese medicine (reviewed in refs. (14–18)). Studies have focused on dozens of species and products. Several researchers have developed and tested methods for identifying products derived from *Panax* spp. (15, 19–23) and *Dendrobium officinale* (24–28). Other studies have investigated methods to authenticate products made from *Angelica sinensis* (29), *Astragalus membranaceus* (30, 31), *Actaea racemosa* (32), *Fritillaria* spp. (33), *Bupleurum* spp. (34, 35), *Euphorbia pekinensis* (36), *Gentiana macrophylla* (37), *Medicago sativa* and *Trifolium pratense* (38), *Phyllanthus* spp. (39, 40), *Stemona* spp. (41, 42), *Swertia mussotii* (43), and *Verbena officinalis* (44). When testing food products, studies concentrate on developing markers that identify unlisted allergen species (45–47), adulterants and mislabeled products (48–50), and genetically modified organisms (46, 51). The strategy of searching for a few predetermined species has also been developed to test for the use of endangered or protected species. The search for protected species is most common for wood products (e.g., (52–55)), but also applied elsewhere (56, 57).

When researchers are interested in developing tools to distinguish a single or a few species, they most often rely on cpDNA spacers, ITS markers, or RAPDs, or a combination of markers (e.g., (19, 58)). In the case of cpDNA and ITS, the technique may be based on sequencing, differences in RFLP patterns, or sequence-specific PCR amplification. Genetic tools that provide more resolution—such as microsatellites—can be used, and are sometimes applied (e.g., (21, 57)). In the case of herbal medications and foods, DNA tools are often used instead of or in addition to traditional chemical analyses (such as HPLC) due to increased speed, power to discriminate species, and the ability to work with degraded or aged samples, or when a specialized chemical analysis has not been developed for a particular species (32).

Researchers often need high-resolution markers that give resolution at the intraspecific level. The goal is generally either to determine the geographic source of a sample (e.g., (59, 60)) or the identity of the variety or cultivar (e.g., (61–64)). Microsatellites are the most commonly used tool for studies of source population or cultivar, but other markers such as AFLPs and RAPDs can also be utilized (65, 66).

In testing commercially available plant products, researchers put less emphasis on investigations that utilize barcoding to search for dozens or hundreds of potential species identifications. Nonetheless, some studies do use techniques that are meant to survey a large number of unrelated candidate plant species using barcoding approaches, especially for herbal medications where alternate ingredients become indistinguishable after processing (31, 40, 67).

Most publications of methods applicable to plant product forensics do not include a survey of the commercially available product of interest; emphasis is put on development and testing of the molecular methods. However, some researchers have surveyed commercially available products. In studies of herbal medication, researchers found anywhere between 8 and 60% of samples may be mislabeled or adulterated (19, 23, 29, 33, 35, 36, 38–42). Some authors have speculated that some cases of mislabeling or adulteration may be due to confusion between morphologically similar and closely related species, instead of deliberate substitution of a medicinal product (29, 40). In other cases, a small number of herbal medicine samples were all found to contain the correct species (32, 38).

A number of studies have reported the application of DNA tools to survey products not related to herbal medication. In one case, published data was presented in court that demonstrated the illegal use of a patented strawberry cultivar (65). In a test of the accuracy of olive oil variety identity, 1 of 16 samples was improperly labeled; the authors speculated it was due to misidentification based on morphology (62). A survey of horticultural suppliers in Japan found that the at least 5 (and up to 10) of 29 suppliers were selling the endangered *Primula sieboldii* originally collected from distant populations, which presents the danger of introducing maladapted genes into already dwindling wild populations (59). Another study found the presence of undeclared cereals in three of ten products sampled, although the levels of unlisted ingredients were low and may have been due to botanical contamination in agriculture (45). Turmeric samples tested in India were mainly composed of a wild, and less potent, relative of the proper cultivated species used for turmeric powder (49). Finally, an investigation of wine barrels in France found that some barrels supposedly made from wood collected in French forests were actually made from oak wood originating in Eastern Europe (53).

1.3. Studies of Illicit Drugs

Forensic DNA methods have been applied to the identification and study of illegal drugs. Most of the published work centers on *Cannabis sativa* (marijuana). DNA has been used to identify the presence of *Cannabis*, and to "individualize" samples (68). Identifying the presence of marijuana can often be done by chemical means, but DNA can be useful when working with very small amounts of plant material. The first markers used to identify *Cannabis* include those based on sequences of ITS (69) and the RFLP of cpDNA spacers (70). For more detailed, intraspecific differentiation of samples, researchers have developed the use of AFLP (71), RAPD (72), and microsatellite (73) markers. These higher resolution markers have been used to look at patterns in *Cannabis* distribution, the role of asexual reproduction in propagation, differentiation between different crop-use and wild types, and genetic structuring at national and global scales (72, 74, 75).

The use of molecular markers has recently been developed to study *Mitragyna speciosa* ("kratom"), another plant used to make products that are illegal in some countries. Researchers have developed methods to identify its presence and to distinguish it from members of the same genus that are not illegal (76), and have been applied to commercially available samples in Japan (77).

1.4. Microsatellites in Plant Forensics

The protocol below is for microsatellite genotyping, though some steps of the protocol can be applied to different DNA markers. The use of microsatellites is not practical for some applications, such as barcoding where the presence and absence of several unrelated species must be determined. However, microsatellites are highly variable, and thus provide the greatest resolution for forensic applications. Microsatellites can be used to make interspecific distinctions, as well as intraspecific distinctions down to the individual level. The codominant, Mendelian inheritance of microsatellite markers allows for accurate calculations of metrics that are useful in plant forensic applications, metrics that are well established for human DNA fingerprinting which also typically employs microsatellites.

2. Materials

Use sterile deionized water where applicable. All materials can be stored at room temperature, unless otherwise noted.

2.1. Sample Storage, Desiccation, and DNA Extraction

1. Desiccating agent: Indicating DRIERITE (W. A. Hammond DRIERITE Co. LTD, Xenia, OH, USA), composed of ~98% $CaSO_4$, ~2% $CoCl_2$. Samples should be stored in desiccators or air-tight zip-top bags with the desiccating agent. Desiccating agent can be stored at room temperature, though it must be in a dry and sealed container. Silica gel can be used as an alternative to DRIERITE.

2. DNeasy Mini-Plant Extraction Kit (Qiagen Inc, Valencia, CA, USA): other DNA extraction techniques will require specialized reagents or kits.

3. Additional 2-mL disposable microcentrifuge tubes.

4. 95% ethanol.

2.2. Gel Electrophoresis

1. Agarose powder (Sigma-Aldrich Co., St. Louis, MO, USA).

2. 5× Tris–borate–EDTA buffer (Sigma-Aldrich Co.). Dilute the Tris–borate–EDTA (TBE) to 1× concentration by mixing one part 5× TBE with four parts water.

3. Ethidium bromide (EtBr) (Sigma-Aldrich Co.).

4. Bromophenol blue sodium salt for electrophoresis (Sigma-Aldrich Co.). Store at 4°C.

5. 100 bp DNA size standard (1 µg/µL; eEnzyme LLC, Montgomery Village, MD, USA). Store at 4°C.

2.3. Polymerase Chain Reaction

1. BioTherm *Taq* DNA Polymerase, and corresponding 10× buffer (eEnzyme LLC). Store at –20°C (see Note 2).

2. 25 mM $MgCl_2$ (Promega Corporation, Madison, WI, USA). Store at –20°C.

3. 100 mM dNTP master mix (Denville Scientific Inc., Metuchen, NJ, USA). Dilute with water to 10 mM in preparation for PCR. Store at –20°C.

4. Bovine serum albumin (BSA), 25 mg/mL. Store at –20°C.

5. If using the method described by Schuelke (78), four different fluorescent dyes are needed: FAM, NED, VIC, and PET (Applied Biosystems, Foster City, CA, USA). Reconstitute the compounds in water to bring the concentration of stock solution to 100 µM. Dilute a portion of each dye further to 10 µM for use in PCR. Store stock and working solutions at –70°C.

6. Single-stranded forward and reverse PCR primers, approximately 20-bp in length (ordered from Integrated DNA Technologies Inc., Coralville, IA, USA). If using the Schuelke method (78), forward primer must include the fluorescent-labeled M13(–21) universal primer, attached to the 5′ end. Add water to bring the concentration to 100 µM stock solution. Dilute samples to 10 µM in preparation for use in PCR. Store stock solution at –70°C, and working solution at –20°C.

<table>
<tr><td>2.4. Resolution
of Microsatellite
Genotypes</td><td>1. Hi-Di Formamide (Applied Biosystems). Store at −20°C.

2. GeneScan 500 LIZ Size Standard (Applied Biosystems). Store at 4°C.</td></tr>
</table>

3. Methods

The number of microsatellite loci included in a study will depend on the variability at the loci (number of alleles, expected and observed heterozygosity) and the objective of the researcher. When distinguishing different species (usually in the same genus) as few as 4 or 5 loci may be sufficient, whereas 12 or more loci may be needed if identifying individual plants. However, before data from microsatellite loci can be collected, PCR primers must be developed to amplify the microsatellite loci to be genotyped.

3.1. Discovery of Microsatellite Loci and Primer Development

1. Search through the primary literature (e.g., *Molecular Ecology Notes*) or GenBank for previously published microsatellite primers that have been developed in the target species. If no results are found for the species in question, expand the search into the entire genus (or perhaps for the subfamily). Microsatellite loci primers are often heterologous across closely related species. If no results are found, a genomic library must be created to screen for microsatellite loci.

2. To create a genomic library for a given species, DNA must be extracted from at least one sample of each target species. Various methods and kits are available for DNA extraction; we have had good, consistent results using Qiagen's DNeasy Mini-Plant Kit. It is probable that RNase will need to be added at some point during or after extraction before proceeding to cloning and sequencing.

3. The library is built by cloning and sequencing stretches of the genome; the rate of successful microsatellites isolation will increase greatly if an enrichment procedure is used (e.g. (79)). New high-throughput DNA sequencing techniques (such as those developed by 454 Life Sciences and Illumina, Inc.) will be increasingly popular for the isolation of microsatellite loci in the future.

4. If using previously published primers, proceed to step 6 and order primers. Otherwise, suitable potential primers for microsatellite loci need to be identified. Use the sequences flanking a repeating motif microsatellite to choose potential primers used to target the locus in question (see Note 1). One primer will be arbitrarily named "forward" and the other "reverse." Each primer should be approximately 20 bp in length.

The primers will have the same sequence in the 5′–3′direction as the regions flanking the microsatellite.

5. When developing potential primers, care should be taken to avoid primers that self-dimerize, may bind to one another, create hairpins, or have a T_m outside of typical ranges acceptable for successful PCR. Various tools exist to prevent these problems. We recommend using two free web-based tools in conjunction: *Primer3* (http://frodo.wi.mit.edu/primer3/) and *OligoCalc* (http://www.basic.northwestern.edu/biotools/oligocalc.html).

6. Once promising forward and reverse primers are chosen, they can be ordered them from any number of DNA technology firms (such as Integrated DNA Technologies). Depending on the method used for resolution of microsatellite genotypes, it may be necessary to order primers with additional nucleotide sequences attached. One common method for genotyping microsatellite genotypes requires the addition of the fluorescent-labeled M13(–21) universal primer, attached to the 5′ end of the forward primer (78).

3.2. Optimization of PCR Conditions

If using newly developed primers, tests for optimal PCR conditions (with respect to reagent concentration and annealing temperature) will be necessary. Reports on previously published primers should include information on reagent concentration and optimal annealing temperature. However, optimization of PCR conditions will likely be useful—and probably necessary if the primers were developed for a different species.

1. In the optimization of PCR conditions, every reaction will have these common characteristics:

   (a) Total reaction volume is 10 μL, achieved by addition of water as necessary.

   (b) Extracted DNA (use ~0.8–2 μL from the DNeasy Mini-Plant Kit extraction). Depending on the quality of DNA extraction and presence of PCR inhibitors that are often present in plant tissue, it may be useful to dilute the sample with water by a factor of 5 or 10.

   (c) 0.05 μL of *Taq* polymerase (final concentration of 0.025 U/μL).

   (d) 0.5 μL dNTP master mix (final concentration of 500 μM).

   (e) 0.6 μL BSA (final concentration of 1.5 μg/μL).

   (f) 0.5 μL forward primer with M13(–21) sequence (final concentration of 0.5 μM).

   (g) 0.6 μL reverse primer (final concentration of 0.6 μM).

(h) For thermal cycling conditions, preheat at 94°C for 5 min. Follow preheating with 35 cycles of denaturing, annealing, and elongation. The duration for each step in the cycle should be 30 s, though longer elongation time will be needed if the region being amplified approaches or is greater than 500 bp in length. Finish with a 5–7-min elongation step. Use 94–95°C for denaturing, and 72°C for elongation. The annealing temperature will be determined after PCR optimization.

2. In the search for optimal PCR conditions, the following factors will vary between reactions:

(a) MgCl$_2$ concentration, varying from 1.5 to 3 mM. We recommend testing a low, moderate, and high concentrations of MgCl$_2$ (see Note 2).

(b) Annealing temperature during thermal cycling. Test annealing temperatures in the range of $T_m \pm 5°C$, where T_m is the lesser melting temperature for the two primers. The most effective way to test the range of annealing temperatures is by utilizing the temperature gradient function on a thermal cycler (see Note 3).

(c) When testing for optimal PCR conditions, we recommend repeating the test for at least two or three different samples of extracted DNA.

(d) Consider testing a variety of dilution levels of the DNA extractions.

3. The layout for a typical PCR plate—using DNA extracted from three different samples, three MgCl$_2$ concentrations, and a thermal cycler with a vertical temperature gradient—is shown in Fig. 1. Create a cycling program as described in (h) of step 1, but use a gradient of annealing temperatures ranging from $T_m - 5°C$ to $T_m + 5°C$.

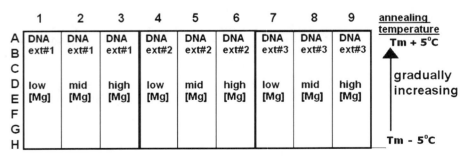

Fig. 1. Suggested arrangement of PCR plate during optimization. The arrangement shown includes DNA extracted from three different samples (ext#1–3), three different MgCl$_2$ concentrations (1.5, 2.25, 3 mM), and a range of annealing temperatures. There are eight different reactions for each combination of DNA extract and MgCl$_2$ concentration (aligned in columns), which differ only in the annealing temperature. The T_m used to calculate the range of annealing temperatures is the lesser melting temperature for the two primers used in the PCR.

4. The success of each reaction can be visualized on an agarose gel, stained with EtBr. To make a 1.25% agarose gel, add 1.25 g of agarose to 100 mL of 1× TBE solution. Slowly add the agarose while stirring to avoid clumps of unsuspended agarose. After adding the agarose, heat on a hot plate while stirring or microwave the mixture to boiling. Add 3 µL of EtBr after cooling for approximately 5 min. Cast the gel, creating wells that will hold at least 5 µL of samples for testing.

5. Add a mixture of 2.5 µL PCR product and 0.5–1 µL of bromophenol blue solution to each well (see Note 4). Load 100-bp DNA ladder (approximately 0.25–0.5 µg) into at least one well.

6. During electrophoresis, run 80–90 V through the gel for 40 min. Use UV light to visualize DNA bands, and print or store a digital image of the gel for later analysis.

7. Use the results of the electrophoresis to choose optimal PCR conditions. Using the 100-bp size standard as a reference, look for amplified DNA in the expected size range for the microsatellite being tested (as determined by the original sequences collected from the genomic library). Aim to reach a proper balance between strength of amplification and PCR specificity. You should expect to see increasing strength of amplification, but decreasing specificity, as the concentration of $MgCl_2$ increases (Fig. 2). The opposite pattern (decreasing strength but increasing specificity) is expected with increasing annealing temperature (Fig. 3).

8. We recommend choosing the lowest concentration of $MgCl_2$ that produces clearly visible bands, in order to strengthen specificity and replication accuracy.

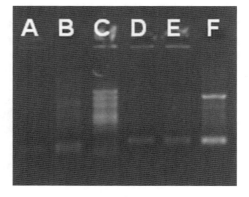

Fig. 2. Effect of $MgCl_2$ concentration during PCR optimization. Columns *A–C* used a higher annealing temperature than columns *D–F*. Note increased amplification—but decreased specificity—in high Mg concentration (3 mM; columns *C* and *E*), and weak amplification at low Mg concentration (1.5 mM; columns *A* and *D*).

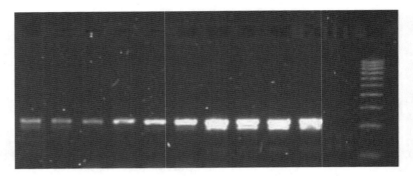

Fig. 3. Gradient of annealing temperature used for PCR optimization of microsatellite primers. In this case, the annealing temperature decreased from *left* to *right*. The far right lane contains 100-bp size standard. Note the increasing degree of amplification, but possible decrease in specificity, with decreased temperature.

9. When choosing the annealing temperature for future PCRs, the highest functioning annealing temperature has highest specificity. However, we recommend an annealing temperature of 1 or 2°C below this temperature to ensure consistent amplification and adequate product (see Note 5).

10. If none of the reactions in the first attempt at optimization are initially successful, or consistently successful when attempting to amplify target loci in a large number of samples, try increasing and decreasing the strength of dilution of the extracted DNA. If problems continue, consider attempting other modifications to the process, redesigning primers, or choosing another locus to optimize (see Note 6).

11. Once optimal PCR conditions (reagent concentrations and annealing temperature) have been chosen and verified, continue with DNA extraction and PCR of each sample for each locus of interest.

3.3. Microsatellite Genotyping and Data Collection

The methods described here are applicable to the Applied Biosystems 3730 DNA Analyzer, a capillary sequencer and fragment analyzer. Other machines and manufacturers will require specialized preparation.

1. For each amplification reaction, mix 1.5 μL of PCR product with 8.2 μL of Hi-Di, and 0.3 μL of LIZ 500 Size Standard (the proportions may change if multiplexing, but total reaction should always be at least 10 μL).

2. Once a plate or half-plate is ready for analysis, incubate it at 95°C for 5 min. Immediately follow with incubation on ice for 5 min.

3. Once the correctly formatted and labeled digital files are prepared and loaded into the software associated with the

3730 DNA Analyzer, the plate is ready for loading and analysis.

4. Data on the length of microsatellite fragments (i.e., genotypes) can be visualized and collected using GeneMapper software (Applied Biosystems; see Note 7).

5. Various computer programs and plug-ins, such as Genepop (80, 81) and GenAlEx (82), can check for complicating issues with the microsatellite genotypes, such as null alleles and linkage disequilibrium. These issues may alter calculations and conclusions associated with forensic analysis.

6. Statistical analyses of microsatellite genotype data that may be useful in forensic applications include assignment tests (83), Bayesian clustering techniques implemented in *Structure* (84), and calculating the PPI. Note that analysis of microsatellite data requires a survey of the potential source populations (or species) in question, especially when calculating the PPI.

4. Notes

1. Repeating motifs of two nucleotides (dinucleotide repeats) tend to be common and often more variable, but can be more difficult to score at later stages. In some cases, allelic stutter and heterozygotes will be difficult to distinguish. When developing microsatellites, preference should be given to tri- and tetranucleotide repeats that can be more accurately scored, unless there are too few highly variable repeat motifs with greater than two nucleotides.

2. MgCl$_2$ is often included in the PCR buffer sold with *Taq* polymerase, and should be included when calculating final concentrations. The 10× buffer listed here (from eEnzyme LLC) contains 15 mM MgCl$_2$.

3. If the option is available, machines with vertical temperature gradients are more useful than those with horizontal temperature gradients. A larger number of recipe-extraction combinations can be tested with vertical temperature gradients, and it is easier to arrange samples in an intuitive pattern during electrophoresis.

4. If your thermal cycling machine utilizes a vertical temperature gradient (as opposed to horizontal), load the PCR optimization samples onto the gel sequentially by column. Carefully keep track of the arrangement of samples on the gel and use the same pattern (e.g., increasing annealing temperature) for each recipe-extraction combination (each column, as arranged in Fig. 1), to make it easier to identify general patterns in amplification success.

5. If primer optimization does not yield successful, consistent amplification, test annealing temperatures lower than initially suggested, such as $T_m - 8$ or $9°C$. Care should be taken to avoid amplification of nontarget regions.

6. If PCR fails, it may be also due to contamination or degradation of reagents. This is especially likely if a previously successful PCR recipe begins to fail consistently. Check the efficacy of reagents by substituting old working stock with samples that are unlikely to be contaminated or adulterated. Single-stranded DNA primers are especially vulnerable to degradation, and should be the first reagents tested. If reagents are not the cause of failure, the literature contains suggestions for other modifications that may increase the likelihood of a successful PCR (85).

7. A common error encountered when using GeneMapper software to resolve microsatellite genotypes is "No sizing data." This usually means that a greater volume of LIZ 500 Size Standard should be used when creating the mixture with PCR product and Hi-Di Formamide.

References

1. Rubinoff, D., Cameron, S., and Will, K. (2006) Are plant DNA barcodes a search for the Holy Grail?, *Trends in Ecology & Evolution 21*, 1–2.

2. Taberlet, P., Coissac, E., Pompanon, F., Gielly, L., Miquel, C., Valentini, A., Vermat, T., Corthier, G., Brochmann, C., and Willerslev, E. (2007) Power and limitations of the chloroplast *trnL* (UAA) intron for plant DNA barcoding, *Nucleic Acids Research 35*.

3. Ferri, G., Alu, M., Corradini, B., and Beduschi, G. (2009) Forensic botany: species identification of botanical trace evidence using a multigene barcoding approach, *International Journal of Legal Medicine 123*, 395–401.

4. Ward, J., Gilmore, S. R., Robertson, J., and Peakall, R. (2009) A grass molecular identification system for forensic botany: A critical evaluation of the strengths and limitations, *Journal of Forensic Sciences 54*, 1254–1260.

5. Mestel, R. (1993) Murder trial features trees genetic fingerprint, *New Scientist 138*, 6.

6. Yoon, C. K. (1993) Forensic Science - Botanical witness for the prosecution, *Science 260*, 894–895.

7. Korpelainen, H., and Virtanen, V. (2003) DNA fingerprinting of mosses, *Journal of forensic sciences 48*, 804–807.

8. Craft, K. J., Owens, J. D., and Ashley, M. V. (2007) Application of plant DNA markers in forensic botany: Genetic comparison of *Quercus* evidence leaves to crime scene trees using microsatellites, *Forensic Science International 165*, 64–70.

9. Lee, C. L., Miller Coyle, H., Carita, E., Ladd, C., Yang, N. C. S., Palmbach, T. M., Hsu, I. C., and Lee, H. C. (2006) DNA analysis of digested tomato seeds in stomach contents, *American Journal of Forensic Medicine and Pathology 27*, 121–125.

10. Lee, C. L., Miller Coyle, H., and Lee, H. C. (2007) Genetic analysis of individual seeds by amplified fragment length polymorphism method, *Croatian Medical Journal 48*, 563–565.

11. Lee, E. J., Kim, S. C., Hwang, I. K., Yang, H. J., Kim, Y. S., Han, M. S., Yang, M. S., and Lee, Y. H. (2009) The identification of ingested dandelion juice in gastric contents of a deceased person by direct sequencing and GC-MS methods, *Journal of Forensic Sciences 54*, 721–727.

12. Techen, N., Crockett, S. L., Khan, I. A., and Scheffler, B. E. (2004) Authentication of medicinal plants using molecular biology techniques to compliment conventional methods, *Current Medicinal Chemistry 11*, 1391–1401.

13. Rachmayanti, Y., Leinemann, L., Gailing, O., and Finkeldey, R. (2009) DNA from processed and unprocessed wood: Factors influencing the isolation success, *Forensic Science International-Genetics 3*, 185–192.

14. Joshi, K., Chavan, P., Warude, D., and Patwardhan, B. (2004) Molecular markers in herbal drug technology, *Current Science 87*, 159–165.

15. Yip, P., Chau, C., Mak, C., and Kwan, H. (2007) DNA methods for identification of Chinese medicinal materials, *Chinese Medicine 2*, 9.

16. Zhang, Y. B., Shaw, P. C., Sze, C. W., Wang, Z. T., and Tong, Y. (2007) Molecular authentication of Chinese herbal materials, *Journal of Food and Drug Analysis 15*, 1–9.

17. Sucher, N. J., and Carles, M. C. (2008) Genome-based approaches to the authentication of medicinal plants, *Planta Medica 74*, 603–623.

18. Jiang, Y., David, B., Tu, P. F., and Barbin, Y. (2010) Recent analytical approaches in quality control of traditional Chinese medicines-A review, *Analytica Chimica Acta 657*, 9–18.

19. Mihalov, J. J., Marderosian, A. D., and Pierce, J. C. (2000) DNA identification of commercial ginseng samples, *Journal of Agricultural and Food Chemistry 48*, 3744–3752.

20. Fushimi, H., Komatsu, K., Isobe, M., and Namba, T. (1996) 18 S ribosomal RNA gene sequences of three *Panax* species and the corresponding Ginseng drugs, *Biological & Pharmaceutical Bulletin 19*, 1530–1532.

21. Qin, J. H., Leung, F. C., Fung, Y. S., Zhu, D. R., and Lin, B. C. (2005) Rapid authentication of ginseng species using microchip electrophoresis with laser-induced fluorescence detection, *Analytical and Bioanalytical Chemistry 381*, 812–819.

22. Ha, W. Y., Shaw, P. C., Liu, J., Yau, F. C. F., and Wang, J. (2002) Authentication of *Panax ginseng* and *Panax quinquefolius* using amplified fragment length polymorphism (AFLP) and directed amplification of minisatellite region DNA (DAMD), *Journal of Agricultural and Food Chemistry 50*, 1871–1875.

23. Del Serrone, P., Attorri, L., Gallinella, B., Gallo, F. R., Federici, E., and Palazzino, G. (2006) Molecular identification of *Panax ginseng* C.A. Meyer in ginseng commercial products, *Natural Product Communications 1*, 1137–1140.

24. Lau, D. T. W., Shaw, P. C., Wang, J., and But, P. P. H. (2001) Authentication of medicinal *Dendrobium* species by the internal transcribed spacer of ribosomal DNA, *Planta Medica 67*, 456–460.

25. Ding, X. Y., Wang, Z. T., Zhou, K. Y., Xu, L. S., Xu, H., and Wang, Y. Q. (2003) Allele-specific primers for diagnostic PCR authentication of *Dendrobium officinale*, *Planta Medica 69*, 587–588.

26. Zhang, Y. B., Wang, J., Wang, Z. T., But, P. P. H., and Shaw, P. C. (2003) DNA microarray for identification of the herb of *Dendrobium* species from Chinese medicinal formulations, *Planta Medica 69*, 1172–1174.

27. Xu, H., Wang, Z. T., Ding, X. Y., Zhou, K. Y., and Xu, L. S. (2006) Differentiation of *Dendrobium* species used as "Huangcao Shihu" by rDNA ITS sequence analysis, *Planta Medica 72*, 89–92.

28. Sze, S. C. W., Zhang, K. Y. B., Shaw, P. C., But, P. P. H., Ng, T. B., and Tong, Y. (2008) A DNA microarray for differentiation of the Chinese medicinal herb *Dendrobium officinale* (Fengdou Shihu) by its 5 S ribosomal DNA intergenic spacer region, *Biotechnology and Applied Biochemistry 49*, 149–154.

29. Feng, T., Liu, S., and He, X. J. (2010) Molecular authentication of the traditional Chinese medicinal plant *Angelica sinensis* based on internal transcribed spacer of nrDNA, *Electronic Journal of Biotechnology 13*.

30. Ma, X. Q., Duan, J. A., Zhu, D. Y., Dong, T. T. X., and Tsim, K. W. K. (2000) Species identification of *Radix Astragali* (Huangqi) by DNA sequence of its 5 S-rRNA spacer domain, *Phytochemistry 54*, 363–368.

31. Guo, H. Y., Wang, W. W., Yang, N., Guo, B. L., Zhang, S., Yang, R. J., Yuan, Y., Yu, J. L., Hu, S. N., Sun, Q. S., and Yu, J. (2010) DNA barcoding provides distinction between Radix Astragali and its adulterants, *Science China-Life Sciences 53*, 992–999.

32. Zerega, N. J. C., Mori, S., Lindqvist, C., Zheng, Q. Y., and Motley, T. J. (2002) Using amplified fragment length polymorphisms (AFLP) to identify black cohosh (*Actaea racemosa*), *Economic Botany 56*, 154–164.

33. Wang, C. Z., Li, P., Ding, J. Y., Peng, X., and Yuan, C. S. (2007) Simultaneous identification of Bulbus Fritillariae cirrhosae using PCR-RFLP analysis, *Phytomedicine 14*, 628–632.

34. Yang, Z. Y., Chao, Z., Huo, K. K., Xie, H., Tian, Z. P., and Pan, S. L. (2007) ITS sequence analysis used for molecular identification of the *Bupleurum* species from northwestern China, *Phytomedicine 14*, 416–422.

35. Lin, W. Y., Chen, L. R., and Lin, T. Y. (2008) Rapid authentication of *Bupleurum* species using an array of immobilized sequence-specific oligonucleotide probes, *Planta Medica 74*, 464–469.

36. Xue, H. G., Zhou, S. D., He, X. J., and Yu, Y. (2007) Molecular authentication of the traditional Chinese medicinal plant *Euphorbia pekinensis*, *Planta Medica 73*, 91–93.

37. Xue, C. Y., Li, D. Z., and Wang, Q. Z. (2008) Identification and quantification of the traditional Chinese medicinal plant *Gentiana macrophylla* using Taqman Real-Time PCR, *Planta Medica 74*, 1842–1845.

38. LeRoy, A., Potter, E., Woo, H. H., Heber, D., and Hirsch, A. M. (2002) Characterization and identification of alfalfa and red clover dietary supplements using a PCR-based method, *Journal of Agricultural and Food Chemistry 50*, 5063–5069.

39. Manissorn, J., Sukrong, S., Ruangrungsi, N., and Mizukami, H. (2010) Molecular Phylogenetic analysis of Phyllanthus species in Thailand and the application of polymerase chain reaction-restriction fragment length polymorphism for *Phyllanthus amarus* Identification, *Biological & Pharmaceutical Bulletin 33*, 1723–1727.

40. Srirama, R., Senthilkumar, U., Sreejayan, N., Ravikanth, G., Gurumurthy, B. R., Shivanna, M. B., Sanjappa, M., Ganeshaiah, K. N., and Shaanker, R. U. (2010) Assessing species admixtures in raw drug trade of *Phyllanthus*, a hepato-protective plant using molecular tools, *Journal of Ethnopharmacology 130*, 208–215.

41. Fan, L. L., Zhu, S., Chen, H. B., Yang, D. H., Cai, S. Q., and Komatsu, K. (2009) Identification of the botanical source of Stemonae Radix based on polymerase chain reaction with specific primers and polymerase chain reaction-restriction fragment length polymorphism, *Biological & Pharmaceutical Bulletin 32*, 1624–1627.

42. Vongsak, B., Kengtong, S., Vajrodaya, S., and Sukrong, S. (2008) Sequencing analysis of the medicinal plant *Stemona tuberosa* and five related species existing in Thailand based on *trn*H-*psb*A chloroplast DNA, *Planta Medica 74*, 1764–1766.

43. Xue, C. Y., Li, D. Z., Lu, J. M., Yang, J. B., and Liu, J. Q. (2006) Molecular authentication of the traditional Tibetan medicinal plant *Swertia mussotii*, *Planta Medica 72*, 1223–1226.

44. Ruzicka, J., Lukas, B., Merza, L., Gohler, I., Abel, G., Popp, M., and Novak, J. (2009) Identification of *Verbena officinalis* based on ITS sequence analysis and RAPD-derived molecular markers, *Planta Medica 75*, 1271–1276.

45. Ronning, S. B., Rudi, K., Berdal, K. G., and Holst-Jensen, A. (2005) Differentiation of important and closely related cereal plant species (Poaceae) in food by hybridization to an oligonucleotide array, *Journal of Agricultural and Food Chemistry 53*, 8874–8880.

46. Mafra, I., Ferreira, I., and Oliveira, M. (2008) Food authentication by PCR-based methods, *European Food Research and Technology 227*, 649–665.

47. Uchino, M., Masubuchi, N., Kurosawa, Y., Noguchi, T., and Takano, K. (2007) Specific primer for detection of wheat, *Journal of the Japanese Society for Food Science and Technology-Nippon Shokuhin Kagaku Kogaku Kaishi 54*, 82–86.

48. Shioda, H., Satoh, K., Nagai, F., Okubo, T., Seto, T., Hamano, T., Kamimura, H., and Kano, I. (2003) Identification of *Aloe* species by random amplified polymorphic DNA (RAPD) analysis, *Journal of the Food Hygienic Society of Japan 44*, 203–207.

49. Sasikumar, B., Syamkumar, S., Remya, R., and Zachariah, T. J. (2004) PCR based detection of adulteration in the market samples of turmeric powder, *Food Biotechnology 18*, 299–306.

50. Marieschi, M., Torelli, A., Poli, F., Bianchi, A., and Bruni, R. (2010) Quality control of commercial Mediterranean oregano: Development of SCAR markers for the detection of the adulterants *Cistus incanus* L., *Rubus caesius* L. and *Rhus coriaria* L, *Food Control 21*, 998–1003.

51. Meyer, R. (1999) Development and application of DNA analytical methods for the detection of GMOs in food, *Food Control 10*, 391–399.

52. White, E., Hunter, J., Dubetz, G., Brost, R., Bratton, A., Edes, S., and Sahota, R. (2000) Microsatellite markers for individual tree genotyping: application in forest crime prosecutions, *Journal of Chemical Technology and Biotechnology 75*, 923–926.

53. Deguilloux, M. F., Pemonge, M. H., and Petit, R. J. (2004) DNA-based control of oak wood geographic origin in the context of the cooperage industry, *Annals of Forest Science 61*, 97–104.

54. Asif, M. J., and Cannon, C. H. (2005) DNA extraction from processed wood: A case study for the identification of an endangered timber species (*Gonystylus bancanus*), *Plant Molecular Biology Reporter 23*, 185–192.

55. Tnah, L. H., Lee, S. L., Ng, K. K. S., Tani, N., Bhassu, S., and Othman, R. Y. (2009) Geographical traceability of an important tropical timber (*Neobalanocarpus heimii*) inferred from chloroplast DNA, *Forest Ecology and Management 258*, 1918–1923.

56. Chen, F., Chan, H. Y. E., Wong, K. L., Wang, J., Yu, M. T., But, P. P. H., and Shaw, P. C. (2008) Authentication of *Saussurea lappa*, and endangered medicinal material, by ITS DNA and 5 S rRNA sequencing, *Planta Medica 74*, 889–892.

57. Eurlings, M. C. M., van Beek, H. H., and Gravendeel, B. (2010) Polymorphic microsatellites for forensic identification of agarwood (*Aquilaria crassna*), *Forensic Science International 197*, 30–34.

58. Kaundun, S. S., and Matsumoto, S. (2004) PCR-based amplicon length polymorphisms (ALPs) at microsatellite loci and indels from non-coding DNA regions of cloned genes as a means of authenticating commercial Japanese green teas, *Journal of the Science of Food and Agriculture 84*, 895–902.

59. Honjo, M., Ueno, S., Tsumura, Y., Handa, T., Washitani, I., and Ohsawa, R. (2008) Tracing the origins of stocks of the endangered species *Primula sieboldii* using nuclear microsatellites and chloroplast DNA, *Conservation Genetics 9*, 1139–1147.

60. Tnah, L. H., Lee, S. L., Ng, K. K. S., Faridah, Q. Z., and Faridah-Hanum, I. (2010) Forensic DNA profiling of tropical timber species in Peninsular Malaysia, *Forest Ecology and Management 259*, 1436–1446.

61. Bandelj, D., Jakse, J., and Javornik, B. (2002) DNA fingerprinting of olive varieties by microsatellite markers, *Food Technology and Biotechnology 40*, 185–190.

62. Pasqualone, A., Montemurro, C., Summo, C., Sabetta, W., Caponio, F., and Blanco, A. (2007) Effectiveness of microsatellite DNA markers in checking the identity of protected designation of origin extra virgin olive oil, *Journal of Agricultural and Food Chemistry 55*, 3857–3862.

63. Brunings, A. M., Moyer, C., Peres, N., and Folta, K. M. (2010) Implementation of simple sequence repeat markers to genotype Florida strawberry varieties, *Euphytica 173*, 63–75.

64. Thomas, M. R., Cain, P., and Scott, N. S. (1994) DNA typing of grapevines: A universal methodology and database for describing cultivars and evaluating genetic relatedness, *Plant Molecular Biology 25*, 939–949.

65. Congiu, L., Chicca, M., Cella, R., Rossi, R., and Bernacchia, G. (2000) The use of random amplified polymorphic DNA (RAPD) markers to identify strawberry varieties: a forensic application, *Molecular Ecology 9*, 229–232.

66. Pafundo, S., Agrimonti, C., Maestri, E., and Marmiroli, N. (2007) Applicability of SCAR markers to food genomics: Olive oil traceability, *Journal of Agricultural and Food Chemistry 55*, 6052–6059.

67. Barthelson, R. A., Sundareshan, P., Galbraith, D. W., and Woosley, R. L. (2006) Development of a comprehensive detection method for medicinal and toxic plant species, *American Journal of Botany 93*, 566–574.

68. Miller Coyle, H., Palmbach, T., Juliano, N., Ladd, C., and Lee, H. C. (2003) An overview of DNA methods for the identification and individualization of marijuana, *Croatian Medical Journal 44*, 315–321.

69. Gigliano, G. S. (1998) Identification of *Cannabis sativa* L. (Cannabaceae) using restriction profiles of the Internal Transcribed Spacer II (ITS2), *Science & Justice 38*, 225–230.

70. Linacre, A., and Thorpe, J. (1998) Detection and identification of cannabis by DNA, *Forensic Science International 91*, 71–76.

71. Miller Coyle, H., Shutler, G., Abrams, S., Hanniman, J., Neylon, S., Ladd, C., Palmbach, T., and Lee, H. C. (2003) A simple DNA extraction method for marijuana samples used in amplified fragment length polymorphism (AFLP) analysis, *Journal of Forensic Sciences 48*, 343–347.

72. Pinarkara, E., Kayis, S. A., Hakki, E. E., and Sag, A. (2009) RAPD analysis of seized marijuana (*Cannabis sativa* L.) in Turkey, *Electronic Journal of Biotechnology 12*.

73. Gilmore, S., Peakall, R., and Robertson, J. (2003) Short tandem repeat (STR) DNA markers are hypervariable and informative in *Cannabis sativa*: implications for forensic investigations, *Forensic Science International 131*, 65–74.

74. Gilmore, S., Peakall, R., and Robertson, J. (2007) Organelle DNA haplotypes reflect crop-use characteristics and geographic origins of *Cannabis sativa*, *Forensic Science International 172*, 179–190.

75. Howard, C., Gilmore, S., Robertson, J., and Peakall, R. (2009) A *Cannabis sativa* STR genotype database for Australian seizures: Forensic applications and limitations, *Journal of Forensic Sciences 54*, 556–563.

76. Sukrong, S., Zhu, S., Ruangrungsi, N., Phadungcharoen, T., Palanuvej, C., and Komatsu, K. (2007) Molecular analysis of the genus *Mitragyna* existing in Thailand based on rDNA ITS sequences and its application to identify a narcotic species: *Mitragyna speciosa*, *Biological & Pharmaceutical Bulletin 30*, 1284–1288.

77. Maruyama, T., Kawamura, M., Kikura-Hanajiri, R., Takayama, H., and Goda, Y. (2009) The botanical origin of kratom (*Mitragyna speciosa*; Rubiaceae) available as abused drugs in the Japanese markets, *Journal of Natural Medicines 63*, 340–344.

78. Schuelke, M. (2000) An economic method for the fluorescent labeling of PCR fragments, *Nature Biotechnology 18*, 233–234.

79. Glenn, T. C., and Schable, N. A. (2005) Isolating microsatellite DNA loci, In *Molecular Evolution: Producing the Biochemical Data, Part B*, pp 202–222.

80. Raymond, M., and Rousset, F. (1995) GENEPOP (version 1.2): Population genetics software for exact tests and ecumenicism, *Journal of Heredity 86*, 248–249.

81. Rousset, F. (2008) GENEPOP'007: a complete re-implementation of the GENEPOP software for Windows and Linux, *Molecular Ecology Resources 8*, 103–106.

82. Peakall, R., and Smouse, P. E. (2006) GENALEX 6: genetic analysis in Excel. Population genetic software for teaching and research, *Molecular Ecology Notes 6*, 288–295.

83. Paetkau, D., Slade, R., Burden, M., and Estoup, A. (2004) Genetic assignment methods for the direct, real-time estimation of migration rate: a simulation-based exploration of accuracy and power, *Molecular Ecology 13*, 55–65.

84. Pritchard, J. K., Stephens, M., Rosenberg, N. A., and Donnelly, P. (2000) Association mapping in structured populations, *American Journal of Human Genetics 67*, 170–181.

85. Radstrom, P., Knutsson, R., Wolffs, P., Lovenklev, M., and Lofstrom, C. (2004) Pre-PCR processing - Strategies to generate PCR-compatible samples, *Molecular Biotechnology 26*, 133–146.

Chapter 5

DNA Purification from Multiple Sources in Plant Research with Homemade Silica Resins

Jian-Feng Li and Jen Sheen

Abstract

DNA purification is a routine procedure in most plant laboratories. Although different kits are available in the market allowing convenient DNA purification, the cumulative cost of purchasing multiple kits for a laboratory can be staggering. Here, we describe a protocol using homemade silicon dioxide matrix for DNA purification from *Escherichia coli* and *Agrobacterium tumefaciens* cells, PCR and restriction digestion mixtures, agarose gel slices and plant tissues. Compared with the commercial kits, this protocol enables easy DNA purification from diverse sources with comparable yield and purity at negligible expenses.

Key words: DNA purification, Silicon dioxide, *Escherichia coli*, *Agrobacterium tumefaciens*, PCR product, Agarose gel, Plant tissue

1. Introduction

Research in plant molecular biology involves DNA purification on a daily basis. In an attempt to develop a versatile and affordable method that could replace expensive commercial kits in diverse-purpose DNA purification procedures, we explored the silica-based technique which takes advantage of the ability of DNA to bind to silica particles in the presence of chaotropic salt (1). In particular, the cheap chemical compound silicon dioxide was used as the DNA binding matrix, and sodium iodide was added to the cell lysate or DNA-containing solution as the chaotropic salt to facilitate DNA binding to the silica matrix. Nonspecifically bound impurities were eliminated by subsequent washing steps and high-quality DNA was readily eluted from the silica particles with as little as 5 μl water. DNA prepared by this protocol could be directly applied to PCR, restriction digestion, DNA sequencing analysis, or transient expression assays through biolistic bombardment or

Nikolaus J. Sucher et al. (eds.), *Plant DNA Fingerprinting and Barcoding: Methods and Protocols*, Methods in Molecular Biology, vol. 862, DOI 10.1007/978-1-61779-609-8_5, © Springer Science+Business Media, LLC 2012

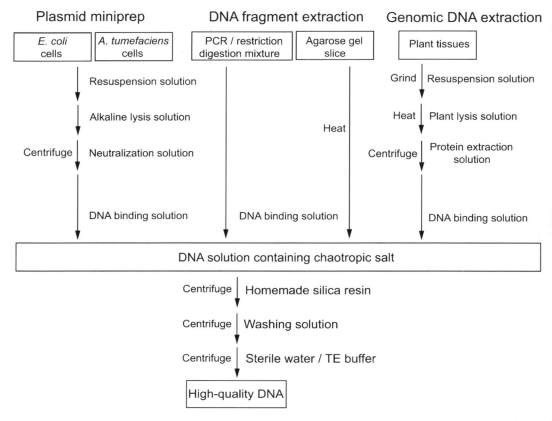

Fig. 1. Flow diagram of the silica protocol for multiple purpose DNA purification in plant research. All procedures are performed at room temperature unless otherwise indicated. The centrifuge step means a 10-s centrifugation at 16,000 × *g*. The heating treatment is conducted at 70°C.

protoplast transfection (2). We have extensively simplified and streamlined this protocol to optimize its time, labor, and cost efficiency for multiple purpose DNA purification (Fig. 1).

2. Materials

All reagents used in this protocol were analytical grade. All solutions were prepared with ultrapure water prepared by purifying deionized water to obtain a sensitivity of 18 MΩ cm at 25°C, and were stored at room temperature unless otherwise stated.

2.1. Bacterial Strain and Growth Medium

1. *Escherichia coli* strain TOP10 or MC1061 and *Agrobacterium tumefaciens* strain GV3101 were used.

2. LB liquid medium: 10 g/l tryptone, 5 g/l yeast extract, and 10 g/l NaCl.

2.2. Plant Tissue	Seven-day-old *Arabidopsis thaliana* (Col-0), *Nicotiana benthamiana* seedlings, and ten-day-old *Zea mays* leaves were used.

2.3. Shared Solutions for Diverse DNA Purification

1. Silica particles: Weigh a 50-ml Falcon tube. Mix 5 g silicon dioxide (Sigma, S5631) with 50 ml sterile water in the Falcon tube and settle the tube upright overnight (see Note 1). Remove the upper fraction containing fine silica particles and resuspend the pellet in 50 ml sterile water. Resettle the tube upright overnight. After discarding the supernatant, weigh the Falcon tube again and calculate the net weight of the remaining silicon dioxide. Resuspend the silica pellet in sterile water to make a final concentration of approximately 100 mg/ml. One milligram of silicon dioxide is able to bind 3–4.5-μg DNA (3). The slurry can be stored at room temperature and be stable for over 12 months.

2. DNA binding solution: 6 M NaI (Sigma, 217638-500G). Filter the solution to remove impurities and store in a dark container at 4°C (see Note 2).

3. Washing solution: 50% (v/v) ethanol, 10 mM Tris–HCl, pH 7.5, 100 mM NaCl, 1 mM EDTA.

4. TE buffer: 10 mM Tris–HCl, pH 8.0, 1 mM EDTA.

2.4. Bacterial Lysis Solutions

1. Resuspension solution: 50 mM Tris–HCl, pH 7.5, 10 mM EDTA, 100 μg/ml RNase A. Store at 4°C.

2. Alkaline lysis solution: 0.2 M NaOH, 1% SDS.

3. Neutralization solution: 1.32 M KOAc. Use acetic acid to adjust the pH value to 4.8.

2.5. Plant DNA Extraction Solutions

1. Plant lysis solution: 10% SDS.

2. Protein extraction solution: Phenol:chloroform:isoamyl alcohol (25:24:1, v/v). Cover the solution with a water phase containing β-mercaptoethanol to prevent the oxidation of phenol. Store at 4°C.

3. Methods

All the procedures were carried out at room temperature unless otherwise stated.

3.1. Plasmid Purification from E. coli or A. tumefaciens by Silica Matrix (see Note 3)

1. Grow 2 ml *E. coli* and *A. tumefaciens* cells at 37°C and 28°C, respectively, until the OD600 of the culture reaches 2.0. Collect the cells in a 2-ml round-bottom microfuge tube by centrifugation at 16,000 × *g* for 30 s.

2. Resuspend the cell pellet in 100 µl resuspension solution by brief vortex.

3. Add 100 µl alkaline lysis solution and invert the microfuge tube for five times (see Note 4).

4. Add 100 µl neutralization solution and invert the tube for five times.

5. Centrifuge the tube at $16,000 \times g$ for 5 min. In the meantime, prepare fresh 1.5-ml microfuge tube and add 500 µl DNA binding solution to each tube.

6. After centrifugation, transfer the supernatant to the prepared 1.5 ml microfuge tube and invert the tube for five times.

7. Add 20 µl silica particles to the tube, mix well, and sit the tube on the rack for 2 min (see Note 5).

8. Pellet the silica particles by a 10-s centrifugation. Pour off the supernatant and gently tap the inverted tube against a pile of Kimwipes to drain the liquid.

9. Wash the silica particles with 500 µl washing solution by vigorous vortex (see Note 6).

10. Repeat steps 8 and 9 (see Note 6).

11. Pellet the silica matrix by a 10-s centrifugation and remove the supernatant by a pipette.

12. Centrifuge for another 10 s and carefully pipette off the residual liquid (see Note 7).

13. Add 40 µl sterile water to resuspend the pellet by brief vortex and place the tube at 70°C for 2 min.

14. Centrifuge the tube at $16,000 \times g$ for 2 min and transfer 37 µl supernatant containing the eluted plasmid DNA to a fresh tube (see Notes 8 and 9).

3.2. DNA Purification from Solution or Agarose Gel Slice

1. Add 150 µl DNA binding solution to up to 50 µl PCR or restriction digestion mixture in a 1.5-ml microfuge tube and invert the tube for five times. For gel purification, add 300 µl DNA binding solution per 100 mg gel slice and heat the microfuge tube at 70°C for 3 min to dissolve the gel (see Note 10).

2. Add 10 µl silica matrix, mix well, and incubate for 2 min.

3. Pellet the matrix by a 10-s centrifugation and remove the supernatant by a pipette.

4. Wash the matrix in 500 µl washing solution by vigorous vortex.

5. Repeat steps 3 and 4.

6. Pellet the matrix by a 10-s centrifugation and discard the supernatant as much as possible.

7. Centrifuge for another 10 s and pipette off the trace amount of liquid (see Note 7).

8. Resuspend the matrix in 5–30 μl sterile water and place the tube at 70°C for 2 min.

9. Centrifuge the tube at 16,000 × g for 2 min and transfer the DNA eluate to a fresh tube (see Note 11).

3.3. Genomic DNA Purification from Plant Tissue

1. Place approximately 10 mg of plant material in a 1.5-ml microfuge tube.

2. Add 200 μl resuspension solution and grind the tissue with a Micro-Grinder homogenizer (Research Products International Corporation) (see Note 12).

3. Add 30 μl plant lysis solution to the homogenate and invert the tube for five times.

4. Place the tube at 70°C for 10 min.

5. Add 250 μl protein extraction solution and vortex the mixture vigorously for 30 s.

6. Centrifuge the tube at 16,000 × g for 5 min at 4°C and transfer the upper phase to a fresh tube (see Note 13).

7. Mix with 500 μl DNA binding solution by inverting the tube for five times.

8. Add 20 μl silica matrix and sit the tube for 2 min (see Note 5).

9. Pellet the matrix by a 10-s centrifugation and remove the supernatant by a pipette.

10. Wash the matrix in 1 ml washing solution by vigorous vortex (see Note 6).

11. Repeat steps 9 and 10 (see Note 6).

12. Pellet the matrix by a 10-s centrifugation and pipette off the supernatant.

13. Centrifuge for another 10 s and remove the residual liquid by a pipette.

14. Add 40 μl sterile water to resuspend the pellet and heat the tube at 70°C for 2 min.

15. Centrifuge at 16,000 × g for 2 min and transfer 37 μl supernatant containing the eluted genomic DNA to a fresh tube (see Note 8).

4. Notes

1. Instead of settling the tube upright overnight, the larger size silica particles can be pelleted by centrifugation at low speed [e.g., 5 min centrifugation at $400 \times g$ using a CL2 centrifuge (Thermo Scientific)].

2. It is normal that the DNA binding solution will slowly turn pale yellow upon storage due to oxidation. The solution is usable within 3 months under the indicated storage conditions.

3. This miniprep protocol can also be followed to purify larger amount of DNA (i.e., midiprep or maxiprep) from the samples by scaling up the input of all reagents accordingly.

4. In case of plasmid purification from *A. tumefaciens*, after adding the alkaline lysis solution, let the microfuge tube sit on the rack at room temperature for 5 min to allow a better lysis of *A. tumefaciens* cells.

5. It is normal to see the precipitation of silica particles during incubation.

6. Generally, 10–15 s vortex at the maximal speed of the vortexer is sufficient. During the vortex, the silica particles may still keep in small patches of matrix and it is unnecessary to completely break the silica patches.

7. A 30-s incubation of the microfuge tube at 70°C with the lid open can guarantee a complete elimination of ethanol in the tube. The ethanol contamination at this step may affect subsequent DNA manipulation and lead to overflow when loading sample for DNA electrophoresis.

8. The remaining 3 μl liquid should be abandoned due to a slight contamination by disturbed silica particles. The pelleted silica particles can diffuse into the supernatant after sitting the tube for over 1 min. When large amount of minipreps (e.g., ten samples) are performed at the same time, collect 4–6 DNA eluate at a time and have another 1 min centrifugation before the second round of eluate collection. In practice, the slight silica contamination appears not to affect subsequent DNA analysis such as DNA sequencing. For protection against DNase digestion or pH fluctuations, TE buffer should be used to elute DNA.

9. DNA binding to the silicon dioxide matrix could be efficiently eluted with as little as 5 μl water/TE buffer after heated at 70°C, thus allowing the miniprep DNA to be harvested in sufficiently high concentrations which are necessary for efficient protoplast transfection (4).

10. It is recommended to tap the microfuge tube frequently to facilitate the gel dissolving.

11. The last 2-μl eluate may be abandoned due to a slight contamination by disturbed silica particles.

12. The amount of RNase A included in the resuspension solution is sufficient to remove RNA contamination from DNA prepared from bacterial cells. However, it may not be enough to clean up all the RNA from plant cells. In this case, more RNase A could be added to the resuspension solution for plant DNA purification.

13. When transfer the upper phase, do not touch the middle layer which contains the denatured proteins. It is recommended to sacrifice some upper phase to avoid taking the middle layer accidentally.

Acknowledgment

This work was supported by NIH R01 grant GM070567 to J.S. and a postdoctoral fellowship award from the MGH fund for medical discovery to J.F.L.

References

1. Boom, R., Sol, C.J., Salimans, M.M., Jansen, C.L., Wertheim-van Dillen, P.M., and Noordaa van der, J. (1990) Rapid and simple method for purification of nucleic acids. *J Clin Microbiol* 28, 495–503.

2. Li, J.F., Li, L., and Sheen, J. (2010) Protocol: a rapid and economical procedure for purification of plasmid or plant DNA with diverse applications in plant biology. *Plant Methods* 6,

3. Boyle, J.S. and Lew, A.M. (1995) An inexpensive alternative to glassmilk for DNA purification. *Trends Genet* 11, 8.

4. Yoo, S.D., Cho, Y.H., and Sheen, J. (2007) Arabidopsis mesophyll protoplasts: a versatile cell system for transient gene expression analysis. *Nat Protoc* 2: 1565–1572.

Chapter 6

Random Amplified Marker Technique for Plants Rich in Polyphenols

Tripta Jhang and Ajit Kumar Shasany

Abstract

More than 10,000 publications using the random amplified polymorphic DNA (RAPD) or related arbitrary marker techniques have been published in two decades of its inception in 1990. Despite extensive use, RAPD technique has also attracted some criticisms, mainly for lack of reproducibility. In the light of its widespread applications, the objective of this chapter is to (1) provide a protocol for RAPD assay, (2) identify the potential factors affecting the optimization of the RAPD assays, and (3) provide proper statistical analysis to avoid false positives. It is suggested that after proper optimization, the RAPD is a reliable, sensitive, and reproducible assay having the potential to detect a wide range of DNA variations. Analyses of the relevant fragments generated in RAPD profile allow not only to identify some of the molecular events implicated in the genomic instability but also to discover genes playing key roles in genetic evolution and gene mapping. RAPD markers will continue to be boon for genetic studies of those organisms where yet no sequence information or scanty information is available.

Key words: Polymerase chain reaction, Molecular marker, Random amplified polymorphic DNA, Arbitrary primers, Agarose gel electrophoresis

1. Introduction

With the invention of polymerase chain reaction (PCR) by Mullis in 1983 and introduction of thermostable DNA polymerases in 1988, DNA fingerprinting techniques has been revolutionized to create a large number of markers in shorter time, with nanogram of template DNA, amenable to automation for the high throughput, robust, and reliable assays. Despite enormous advances in molecular marker techniques that have been used to obtain genetic data, random amplified polymorphic DNA (RAPD) remains a useful approach for several reasons (1–6). The method has considerable advantages over other kinds of DNA marker analysis because

Nikolaus J. Sucher et al. (eds.), *Plant DNA Fingerprinting and Barcoding: Methods and Protocols*, Methods in Molecular Biology, vol. 862, DOI 10.1007/978-1-61779-609-8_6, © Springer Science+Business Media, LLC 2012

it is fast, requires little amount of DNA, is suitable for work on anonymous genomes (7), is economic, does not involve complex procedures of DNA sequencing, requires only a simple laboratory with minimum equipment to perform PCR, and can detect polymorphisms in any kind of sequences. In RAPD, DNA fingerprints that incorporate hundreds of polymorphic markers can be produced relatively easily with little or no prior information about the genetics of the organism (8). RAPD fingerprinting is straightforward and can be optimized for reproducibility. The versatility of this technique stems from the simplicity of the information provided by the fingerprint, the high throughput of the method, and the ease with which fingerprints can be generated. RAPD fingerprinting is a PCR-based technique which utilizes a single, arbitrarily chosen short oligonucleotide primer to amplify genome segments flanked by two complimentary primer binding sites in inverted orientation (1). Typically, in standard RAPD analysis, each primer is a 10-mer oligonucleotide with 60–70% G+C content and with no self-complementary ends. Short primers with an arbitrary sequence can be complimentary to a number of sites in a genome. If the sites occur on opposite strands of segment of DNA in inverted orientation and the distance between the sites are short (<5 kbp) enough for PCR, then the segment flanked by the sites can be amplified (see examples illustrated in Figs. 1 and 2). Amplifications on different segments are independent. The primers will hybridize to binding site that are identical or highly homologous to their template nucleotide sequences, although some mismatches especially at the 5′ end are allowed. The number of fragments that can be expected theoretically from one primer annealing with 100% homology (b) can be calculated from primer length (n), target genome size per haploid genome (N), and maxi-

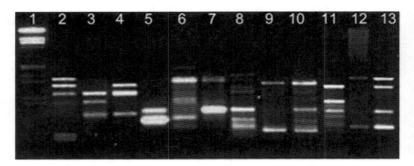

Fig. 1. Coinheritance pattern in *Mentha spicata*. *Lane* 1: Marker lambda Hind III + ECo R1. *Lanes* 2–4: RAPD profiles of *M. spicata* CIMAP/C30, *M. spicata* CIMAP/C33, and the hybrid, respectively, with OPJ 05. *Lanes* 5–7: Profiles with OPJ 14. *Lanes* 8–10: Profiles with OPO 19. *Lanes* 11–13: Profiles with OPT 09.

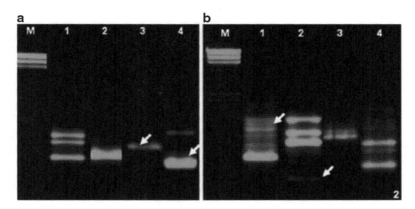

Fig. 2. Amplified profiles of *Phyllanthus* with MAP 09 (**a**) and MAP 10 (**b**) primers from pooled DNA of ten accessions of each species. The *arrows* in *lanes* 3 and 4 of (**a**) indicate the unique fragments from *Phyllanthus debilis* (980 bp) and *Phyllanthus urinaria* (550 bp), respectively. Similarly, the *arrows* in *lanes* 1 and 2 of (**b**) indicate the unique fragments from *Phyllanthus amarus* (1,150 bp) and *Phyllanthus fraternus* (317 bp, light fragment), respectively. M:DNA Marker λ Hind III digest; *Lanes* 1–4: *P. amarus, P. fraternus, P-.debilis, P. urinaria.*

mum fragment length that can be amplified (*f*) $b = 2fN/16n$ (9). Most polymorphism is detected as the presence vs. absence of fragments and may be caused either by failure to prime a site in some individuals because of nucleotide sequence differences or by insertions or deletions in the fragment between two priming sites. Since the process described here uses primers of arbitrary nucleotide sequence to access random segments of genomic DNA to reveal polymorphisms, (1) these are termed as RAPD markers. Nearly all RAPD markers are dominant as DNA segments of the same length are amplified from one individual but not from another. It is not possible to distinguish whether a DNA segment is amplified from a locus that is heterozygous (one copy) or homozygous (two copies) with a dominant RAPD marker. Codominant RAPD markers observed as different sized DNA segments amplified from the same locus are detected only rarely (9, 10). Polymorphism of amplified fragments are mainly due to base substitutions or deletions in the priming sites, insertions that render priming sites too distant to support amplification, or insertions or deletions that change the size of the amplified fragment (1). Though RAPD amplified fragment number is independent of genome size and ploidy level, the amplification reaction is determined in part by competition for genomic priming sites. Primers will preferably bind to target sites with a higher degree of homology. These are more likely available in a more complex genome (9). Southern blot hybridizations experiments by Hallden et al. (11) demonstrated no correlation between target DNA sequence copy number and competitive strength of a fragment (actual DNA sequence). Genetic background is a strong determinant of whether a particular RAPD fragment is amplified or not, as the synthesis of one particular RAPD product is suppressed by the presence of the template for another product in the reaction (12). Eight nucleotides closet to the 3′end of a primer are crucial for the generation of a particular fragment and with decrease in primer length, number of products decreased with an increase in mean size of amplified fragments (13). According to Caetano-Anollés et al (13) DNA amplification is modulated at two levels. First, primer target sites are selected in a template screening phase. The selectivity at this stage is determined by primer sequences and influenced by reaction conditions. Homologous and mismatch annealing may occur, resulting in a complex family of primary amplification products. For an efficient amplification of a given fragment, hairpin loop formation must be out-competed by the primer-template duplex. Since large hairpin loops are weak competitors, very short primers result in longer amplified products. Later several additional types of artifactual inter- and intrastrand interaction can take place, including nested primer annealing to internal regions of RAPD fragments. The brightness of a given fragments will depend on several factors, including the degree of repetitiveness of the targeted DNA region, the extent of primer-template mismatch, and the presence or absence of competing target regions in the genome.

There are reports of non-Mendelian segregation of RAPD fragments (14–17). This includes the presence of nonparental fragments in the progeny where artifactual heteroduplex molecules are formed when two allelic DNA segments differing by one or more base substitutions, insertions, and/or deletions are amplified during the PCR or interaction of PCR products from different loci, because of conformational changes. Conformational changes are caused by nucleotide divergence between the alleles in heteroduplexes which generally migrate at different rates as compared to homoduplexes in gel electrophoresis. Such profile is expected more in outcrosser's than in selfer's due to higher degree of heterozygosity. Others are absence of parental fragments in the progeny as a consequence of competition for target sequences (11, 18) and uniparental inheritance of fragments which originate from organellar DNA (19). A protocol is presented here specially for the medicinal and aromatic plants or crops rich in polyphenols (20) with a discussion on the influence of each reaction component and conditions to optimize the reproducibility and robustness along with better interpretation, escaping false positives.

2. Materials

2.1. Reagents and Chemicals Required

2.1.1. DNA Isolation

Plant Samples for DNA Isolation

Samples of plant material can be collected in the form of leaves, stem, whole flowers, or their parts. Genomic DNA can be extracted from both fresh and dry samples of each material.

1. Tris–Cl pH 8.0 (1.0 M); EDTA pH 8.0 (0.5 M); NaCl (5.0 M); CTAB (20%); chloroform:isoamyl alcohol (24:1 v/v); polyvinylpyrrolidone and β-mercaptoethanol.
2. Extraction buffer: 100 mM Tris–Cl (pH 8.0), 25 mM EDTA, 1.5 M NaCl, 2.5% CTAB, 0.2% b-mercaptoethanol (v/v) (added immediately before use), and 1% PVP (w/v) (added immediately before use).
3. High salt TE buffer: 1 M NaCl, 10 mM Tris–Cl (pH 8.0), and 1 mM EDTA.
4. Liquid nitrogen.
5. Chilled mortar and pestle.

2.1.2. PCR Amplification.

1. Genomic DNA: 12.5–25 ng/μl (see Note 1).
2. Primers: 5 pmol of decanucleotide primers. Custom decanucleotide primers (see Note 2).
3. *Taq* DNA Polymerase (0.2–1 U/μl) supplied in 10 mM Tris–HCl (pH 7.4), 100 mM KCl, 0.1 mM EDTA, 1 mM dithiothreitol, 0.5% Tween20, 0.5% NP-40, and 50% glycerol(see Note 3).
4. *Taq* DNA polymerase buffer (100 mM Tris–HCl (pH 8.3), 500 mM KCl).
5. Thermal Cycler (see Note 4).
6. $MgCl_2$ (15 mM) (see Note 5).
7. Nucleotides: 200 μM of each dNTPs (G, A, T, C).
8. Loading dye: 50% glycerol, 1% SDS, 0.25% bromophenol blue, 0.25% xylene cyanol.
9. Gel staining solution (ethidium bromide 10 mg/ml).
10. Agarose (Sigma chemicals).
11. Gel electrophoresis unit along with power pack.

3. Methods

3.1. DNA Isolation

1. Grind the plant material in liquid nitrogen (3 g fresh tissue or 0.5 g of dry tissue) in prechilled mortar pestle.

2. Transfer the material to 10-ml polypropylene tube and add 3 ml of freshly prepared extraction buffer, mix by inversion to a slurry.

3. Incubate at 60°C in a shaking water bath (100 rpm) for 1–2 h (dry samples may require overnight incubation at 37°C).

4. Add 3 ml of chloroform:isoamylalcohol (24:1) and mix by inversion for about 15 min.

5. Spin at 3,5000 × g for 10 min at 25–30°C.

6. Carefully transfer the upper clear aqueous layer to another 10-ml polypropylene tube.

7. Add 1.5 ml of 5 M NaCl and mix properly (do not vortex) (see Note 6).

8. Add 0.6 volume of isopropanol and let the mixture stand at room temperature for 1 h. After 1 h, slow and careful mixing will produce fibrous nucleic acid that can be scooped and transferred to a 1.5-ml microfuge tube and centrifuged. Alternatively, after mixing with isopropanol, the samples can be centrifuged at 5,6000 × g for 10 min at 25–30°C.

9. Discard the supernatant and wash the pellet with 80% ethanol. Dry the pellet in a vacuum for 15 min and dissolve it in 0.5 ml of high salt TE buffer.

10. Add 5 ml of RNase A and incubate at 37°C for 30 min.

11. Extract with equal volume of chloroform:isoamyl alcohol (24:1).

12. Transfer the aqueous layer to a fresh 1.5-ml microfuge tube and add two volumes of cold ethanol.

13. Spin at 5,6000 × g for 10 min at 25–30°C.

14. Wash the pellet with 80% ethanol.

15. Dry the pellet in a vacuum and dissolve in 200 ml of sterile double distilled water.

16. DNA concentrations can be measured by running aliquots (2 μl) of DNA from each sample loaded on a 0.8% agarose gel to check the quality or by taking the absorbance at 260 nm by Nanodrop.

17. Use about 12.5–25 ng for PCR amplification.

3.2. PCR Amplification

1. All components are thawed on ice.

2. Spin all the reaction components thoroughly for 30 s.

3. In thin, flexible-walled PCR tubes amplification reaction is carried out in a 25-µl volume containing 12.5–25 ng of template DNA.

4. For N number of genotypes, reaction master mix is prepared in a 1.5-ml microcentrifuge tube containing $(N+1)$ × times 0.2 units of *Taq* DNA polymerase, 100 mM each of dNTPs, 1.5 mM $MgCl_2$, and 5 pmol of decanucleotide primers. A negative control using water in place of DNA is recommended (see Note 7). Use of separately designated pipet set for PCR reactions is suggested. If the sample size is large, >60, use of PCR plate is recommended.

5. Mix the reaction master mix properly by centrifuging for 2 min.

6. Aliquot master mix equally (Total/Nµl) in PCR tubes.

7. Spin PCR tubes and immediately keep into a DNA Engine thermal cycler (MJ Research, USA) preset for the desired program.

8. A standardized RAPD program is 4 min at 94°C (initial denaturation step) 40 cycles each consisting of 1 min at 94°C (denaturation), 1 min at 37°C (annealing) (see Note 8), 2 min at 72°C (elongation) followed by 7 min 72°C (final extension), and 4°C HOLD.

3.3. Amplicon Visualization, Scoring, and Analysis

1. Mix amplified products (10 µl) with 3 µl of 6× loading buffer (equal ratio of bromophenol blue and xylene cyanol FF dissolved in glycerol buffer) and load on 1.4% agarose gel containing 5 µg/ml of ethidium bromide (see Note 9).

2. Gel electrophoresis is carried out with 50 V/cm for 3 h and photographed by a gel documentation system.

3. Score RAPD fragments as absent (0) or present (1). When present assess monomorphic or polymorphic. If fragment is polymorphic assess for "repeatability," "heritability," and "transmitability" (see Notes 10 and 11).

4. Exclude low-intensity fragments (see Note 12).

5. Remove fragments of either very high or very low molecular weight (see Note 13).

6. Fragments that are too close together (see Note 14).

7. To analyze RAPD marker data, three coefficients, such as simple matching coefficient, Jaccard's coefficient, and Nei and Li's coefficient are generally used to measure genetic similarities (see Note 15).

8. To find the smallest number of individuals that contain all common fragments, repeatability assay based on a sequential search can be performed starting with the individual that pos-

sesses the most fragments and selecting it and moving to the next individual containing the largest number of fragments that are not found in the first individual, maximizing the complementarities between them. The procedure continues until all individuals are selected, forming a solution of length k (number of individuals) (see Note 16).

9. Binary data generated from reproducible repetitive fragment can be analyzed using Nei and Li coefficient using UPGMA clustering to generate phylogenetic tree by various softwares such as NTSys, DARwin, or Powermarker. Phylogenetic tree so generated can be bootstrapped for the clustering pattern using large iterations using appropriate softwares (see ref. 37).

4. Notes

1. Genomic DNA (12.5–25 ng) with spectrophotometric 260:280 ratios between 1.7 and 1.9 is considered to be pure and of high quality. DNA extractions can be as per (see ref. 20) or through the commercial DNA extraction kits such as Pure Gene™ DNA isolation kit (Gentra Systems Inc.), Nucleon Phytopure (Amershem Biosciences) Geneclean (Qbiogene), Nuc Prep (Applied Biosystems), Aqua Pure DNA Tissue Kit (Bio-Rad), Dynabeads DNA Direct (Dynal), Plant DNazol reagent and Easy DNA (Invitrogen), NucleoSpin, Machrey-Nagel, DNeasy, (Qiagen) etc. following the instructions of the manufacturer. Inhibitory effects of polyphenols, polysaccharides, and RNA on amplification can be reduced by purification of genomic DNA. Inhibition by acidic polysaccharides can also be neutralized by adding Tween-20, dimethysulfoxide (DMSO). or PEG 400 in the reaction mixture. Amplification is inhibited if the template yield is very high or very low.

2. Primers of random base composition, 10-bp long with 40–60 GC% content, with an annealing temperature between 32 and 37°C can be used. Oligonucleotide primers can be purchased from Operon Technologies Inc. (Alameda, CA); University of British Columbia Arbitrary 10-mer kits or can be obtained from the Genosys Biotechnologies Inc. They can also be designed by various primer designing software tools avoiding complementary ends, complete palindromes, self homology, interprimer homology, and GC repeats.

3. Most commercially available normal DNA polymerase for normal amplification have 5′–3′ polymerase activity and 5′–3′ exonuclease activity (primer chewing activity) but lack 3′–5′ exonuclease activity (proof reading activity). Specialized DNA polymerases with proof reading activity are used for correct

base pair matching resulting in enhanced reproducibility of small fragments (*Ampli Taq, Pfu, Phusion, Vent, deep Vent, Klentaql*). By using specific *Taq* polymerases amplification products more than 10 kb can be obtained under optimal conditions.

4. It is being reported that with similar reaction components, RAPD program results in different amplification patterns when used on different thermal cyclers, which can be due to different temperatures of the wells. This indicated different temperature profiles inside tubes, and this must be similar if laboratories are to achieve comparable results. Highly rich AT sequences (genes/genomes) may require reduced extension temperatures (60 or 65°C instead of 72°C) and longer extension times. Slow transition time (ramp rate) between annealing and elongation step temperatures in RAPD is recommended. These studies indicate that reproducible polymorphisms can be generated when standard reaction conditions are used permitting the amplification of appropriate size fragments.

5. 1.5–2.0 mM concentration of magnesium is optimal for DNA polymerase. This depends on template DNA, buffer, and dNTPs and each has the potential to chelate magnesium. If (Mg^{2+}) is too low, no PCR product is seen. If (Mg^{2+}) is too high, undesired PCR products are detected. Optimization of amplification is required by supplementing magnesium concentration in 0.5 mM increments upto 4 mM.

6. This high salt concentration helps in precipitation of polyphenols.

7. A variety of PCR additives and enhancing agents have been used to increase the yield, specificity, and consistency of PCR reactions. While these additives may have beneficial effects on some amplification, it is impossible to predict which agents will be useful in a particular context and, therefore, they must be empirically tested for each combination of template and primers. Amplification specificity and yield can be improved by use of various additives in the PCR such as 0.1% bovine serum albumin (BSA) that has proven particularly useful when attempting to amplify ancient DNA or templates which contain PCR inhibitors such as melanin (see ref. 21). DMSO at 2–10% reduces secondary structures and is particularly useful for GC-rich templates and Betaines or Betaine (mono) hydrate at 1.0–1.7 M for GC-rich templates. Formamide is generally used at 1–5% (see ref. 22). Nonionic detergents stabilize *Taq* polymerase and may also suppress the formation of secondary structure. 0.1–1% Triton X-100, Tween-20, or NP-40 may increase yield and may also increase nonspecific amplification. As little as 0.01% SDS contamination of the template DNA (left-over from the extraction procedure) can inhibit PCR by

reducing *Taq* polymerase activity to as low as 10%, however, inclusion of 0.5% Tween-20 or -40 will effectively neutralize this effect (see ref. 22). TMAC (tetramethyl ammonium chloride) is generally used at a final concentration of 15–100 mM to eliminate nonspecific priming. TMAC is also used to reduce potential DNA–RNA mismatch and improve the stringency of hybridization reactions (see ref. 23). The base analogue 7-deaza-2′-deoxyguanosine may facilitate amplification of templates with stable secondary structures when used in place of dGTP in a ratio of 3:1, 7-deaza-2′-deoxyguanosine:dGTP (see ref. 24).

8. Annealing temperature depends on the primer length, sequence, and concentration. Amplicon melting temperature can be calculated as $T_m = 2°C$ for each AT-pair and 4°C for each GC-pair see ref. 25. The annealing temperature is usually set to about 5°C below T_m. Low annealing temperature typically from 32 to 37°C is used for RAPD primers to allow a certain extent of primer-template mismatching. Exact annealing temperature can be optimized by using Gradient Cycler.

9. Amplified products can be saved for a month at −20°C, but for best resolution it is recommended to load within 24 h.

10. Artifactual fragments may arise from heteroduplex formation between amplified products (see ref. 15) or from other secondary artifacts (see refs. 7, 16). These problems can be partially overcome by rigid laboratory protocols and by performing repeatability tests. Because of the basic design of the technique, it can potentially generate spurious fragments (see refs. 26, 27) and a strict control of working conditions is demanded. Furthermore, due to the anonymous character of polymorphic fragments and the difficulties for establishing homologies, it is also recommended to confine RAPD uses to the specific or intraspecific levels. Comparisons based on genetic distance calculations are accepted provided they do not require parsimony analysis methods. By using stringent protocol conditions and replications, only the fragments that appear repetitively should be scored for further analysis.

11. Assess as monomorphic (found in all individuals) or polymorphic (segregating, if studied in a structured/mapping population). Segregating fragments are further classified as "repeatable," if the fragments are amplified for all parental samples; "heritable," if, for each progeny sample expressing the fragment, at least one of its parents also has the same size of fragment; and "transmittable," if, for each parental sample expressing the fragment, at least one of its six progeny also expressed the fragments. A fragment that failed any of these three criteria in any of the samples was scored "not reproduc-

ible." Repeatability, heritability, and transmitability of finger-print fragments (28) are assessed in relation to the intensity, size, and isolation of the fragments. Band repeatability is defined as: $R_b = 2b_{12}/(b_1 + b_2)$, where b_{12} is the number of individuals possessing fragments b in both replicates and b_1 and b_2 are the numbers of individuals possessing that bands in the first and the second replicate, respectively. R_b can take values between 1 and 0. Mean observed frequency of each bands (f_b) is calculated as $f_b = (b_1 + b_2)/2N$, N being the number of individuals tested. The observed frequency of a band in the population depends on the real frequency of the corresponding priming sites and the repeatability. When no true polymorphism exists for a particular locus, the bands corresponding to the dominant allele can be amplified with equal probability from any individual and then observed frequency reaches its maximum value which is equal to repeatability. Band frequency is plotted against fragments repeatability for each species and annealing temperature. Points in such a plot fit a regression line passing through the origin and with slope equal to one if no polymorphism exists. Positive deviations from this line are error deviations. Points falling significantly below this line indicate that there is true genetic polymorphism, that is, the fragments are not amplified with the same success from any individual (see ref. 29).

12. Fragments intensity is measured as the average intensity of the fragments across all lanes in which it was scored, and total lane intensity was calculated as the sum of the average intensities for all fragments.

13. Amplification of high molecular weight RAPD fragments may be more strongly influenced by differences in DNA quality or PCR reaction conditions, causing them to be less reproducible. Additionally, very low molecular weight fragments may include the so-called primer dimers, or artifactual amplification products produced from the interactions of primers alone.

14. These might result in confusion during fragments scoring, so fragments that were very close in size were removed from the data set. In this case, fragments were excluded if for any two fragments x and y, with x larger than y, log(size of fragments x) – log(size of fragments y) < 0.01.

15. The similarity index is defined as the fraction of shared fragments. For individuals x and y, this is the number of common fragments in their RAPD profiles (n_{xy}) divided by the average number of fragments scored for both individuals: $S_{xy} = 2n_{xy}/(n_x + n_y)$. Autosimilarity ($S_{xx}$) can also be defined as the similarity of an individual with itself in two replicates. Mean autosimilarity and two mean similarity coefficients, one between individuals within the same replicate and the other between

individuals run in different replicates, are calculated. Errors of estimates are based on the between-primer variance of coefficients. Autosimilarity ranges between one, if fragments are completely reproducible, and a value equal to similarity, if differences between individuals are solely the result of error. The magnitude of error is estimated by the proportion of differences between individuals (dissimilarity index), contributed by differences between replicates: $(1 - S_{xx})/(1 - S_{xy})$ (see ref. 30). Artifacts cause bias (underestimation or overestimation) in the computed values of all three similarity coefficients, but Nei and Li's coefficient is affected less by artifacts than the other two. When there are replicate runs, it is easier to estimate s as the sum of the estimated similarity caused by shared positive fragments, $s\,xp$ plus the estimated similarity caused by shared negative fragments, $sx(1 - p)$. The products are summed over all fragments and divided by the total number of fragments. The result is an estimate of $s\,xp$ (see ref. 31).

16. Simulated annealing is a global optimization meta-heuristic method that, starting with a random configuration of objects (i.e., individuals), probabilistically decides to add or delete individuals from this configuration and iterates the procedure minimizing a given quantity (in this case, the loss in number of fragments) (see refs. 32, 33). SITES software (see ref. 34) is used in reserve design procedure to find the smallest number of individuals that (when combined) contain all fragments. HICKORY software can also be used (which estimates the genetic parameters from dominant markers using the Bayesian approach calculating the FST divergence, inbreeding coefficient f, and the genetic diversity Hs) (see ref. 35) and POPGENE (for estimating the proportion of polymorphic loci (P%), Nei's genetic diversity (He), and Shannon genetic diversity (S), assuming Hardy–Weinberg equilibrium) see ref. (36) softwares on an unreduced data set and on data sets in which fragments were eliminated based on repeatability of individuals selected by simulated annealing . Thus, at least in part, the problems of repeatability attributed to RAPD markers, in terms of variation in estimates of genetic parameters when using all loci and only those with high repeatability, can be due to bias in the selection of loci and primers and not necessarily to RAPD per se.

Acknowledgments

We thankfully acknowledge the encouragement of Director, CIMAP and constant financial support by Council of Scientific and Industrial Research and Department of Biotechnology. The help of co-authors in our cited work is also acknowledged.

References

1. Williams JGK, Kubelik AR, Livak KJ, Rafalski JA, Tingey SV(1990)DNA polymorphisms amplified by arbitrary primers are useful as genetic markers. Nucl Acids Res 18:6531–35

2. Lacerda DR, Acedo MDP, Filho JPL, Lovato MB (2002). A técnica de RAPD: uma ferramenta molecular em estudos de conservação de plantas. *Lundiana* 3:87–92.

3. Magalhães M, Martinez RA, Gaiotto FA (2007). Genetic diversity of *Litopenaeus vannamei* cultivated in Bahia State, Brazil. Pesq. Agropec. Bras. 42:1131–1136.

4. Brahmane MP, Mitra K and Mishra SS (2008). RAPD fingerprinting of the ornamental fish *Badis badis* (Hamilton 1822) and *Dario dario* (Kullander and Britz, 2002) (Perciformes, Badidae) from West Bengal, India. Genet. Mol. Biol. 31:789–792.

5. Dutra NC, Telles MP, Dutra DL,Silva Junior NJ (2008). Genetic diversity in populations of the viper *Bothrops moojeni* Hoge, 1966 in Central Brazil using RAPD markers. *Genet. Mol. Res.* 7:603–613

6. Soares TN, Chaves LJ, de Campos Telles MP, Diniz-Filho JA, et al. (2008). Landscape conservation genetics of *Dipteryx alata* ("baru" tree: Fabaceae) from Cerrado region of central Brazil. Genetica 132:9–19.

7. Hadrys H, Balick M, Schierwater B (1992) Applications of random amplified polymorphic DNA (RAPD) in molecular ecology. Molecular Ecology 7:55–63.

8. Bagley M J, Anderson SL, May B (2001) Choice of methodology for assessing genetic impacts of environmental stressors: polymorphism and reproducibility of RAPD and AFLP fingerprints. Ecotoxicology 10:239–244

9. Williams JGK, Hanafey MK, Rafalski JA, Tingey SV(1993). Genetic analysis using random amplified polymorphic DNA markers. Methods Enzymol 218:704–40.

10. Fritsch P, Riseberg L H (1992) High outcrossing rates maintain male and hermaphrodite individuals in populations of the flowering plant. Datisca glomerata Nature 359:633–36

11. Hallden C, Hansen M, Nilsson NO, Hjerdin A, Sall T (1996) Competition as a source of errors in RAPD analysis. Theor Appl Genet 93:1185–1192

12. Reineke A, Karlovsky P, Zebitz CPW (1999) Suppression of randomly primed polymerase chain reaction products(random amplified polymorphic DNA in heterozygous diploids. Mol Ecol 8:1449–55.

13. Caetano-Anollés G, Bassam BJ, Gresshoff PM (1992) DNA fingerprinting MAAPing out a RAPD redefinition?BI/Technology 10:937

14. Ayliffe MA, Lawrence GJ, Ellis JG, Pryor AJ (1994) Heteroduplex molecules formed between allelic sequences cause nonparental RAPD bands. Nucleic Acids Res. 22: 1632–36.

15. Hunt GJ, Page Jr RE (1992). Patterns of inheritance with RAPD molecular markers reveal novel types of polymorphism in the honey bee. Theor. Appl.Genet. 85:15–20.

16. Riedy MF, Hamilton III WJ, Aquadro CF (1992) Excess of non-parental bands in offspring from known primate pedigrees assayed using RAPD PCR. Nucl Acids Res. 20:918.

17. Scott MP, Haymes KM, Williams SM (1992) Parentage analysis using RAPD PCR Nucleic Acid Research20:5493.

18. Heun M, Helentjaris T (1993) Inheritance of RAPDs in F₁ hybrids of corn. Theor. Appl. Genet. 85:961–968.

19. Aagard JE, Vollmer SS, Sorenson FC, Strauss SH (1995) Mitochondrial DNA products among RAPD profiles are frequent and strongly differentiated between races of Douglas-fir Mol Ecol4:441–447.

20. Khanuja SPS, Shasany AK, Darokar MP, Kumar S(1999) Rapid isolation of DNA from dry and fresh samples of plants producing large amounts of secondary metabolites and essential oils. Plant Molecular Biology Reporter 17:1–7

21. Levi A, Rowland LJ, Hartung JS (1993) Production of reliable randomly amplified polymorphic DNA(RAPD) markers from DNA of woody plants.HortScience36:1096–101.

22. Gelfand D H (1988) In Erlich, H. A. (ed.) *PCR Technology*. p.17. Stockton Press, N

23. Wood WI, J GitschierJ, Lasky LA, Lawn RM (1985) Base composition-independent hybridization in tetramethylammonium chloride: a method for oligonucleotide screening of highly complex gene libraries. Proc Natl Acad Sci U S A. 82(6):1585–88

24. Seela F, Driller H (1989) Alternating d(G-C)3 and d(C-G)3 hexanucleotides containing 7-deaza-20-deoxyguanosine or 8-aza-7-deaza-20-deoxyguanosine in place of dG. Nucleic Acids Res.17:901–10.

25. Thein SL, Wallace B (1986)The use of synthetic oligonucleotides as specific hybridization probes in the diagnosis of genetic disorders in Human Genetic Diseases-A Practical approach, Davies KE Ed, IRL Press Oxford UKpp33-50

26. Pérez T, Albornoz J,Domínguez A (1998) An evaluation of RAPD fragment reproducibility and nature. Molecular Ecology 7:1347–57

27. Rabouam C, Comes AM, Bretagnolle V, Humbert JF, et al. (1999). Features of DNA fragments obtained by random amplified polymorphic DNA (RAPD) assays. Mol. Ecol. 8: 493–503.

28. Pérez T, Albornoz J, Domínguez A (1998) An evaluation of RAPD fragment reproducibility and nature. Molecular Ecology7:1347-1357

29. Lynch M (1990) The similarity index and DNA fingerprinting. *Molecular Biology and Evolution*, 7, 478–484.

30. Lamboy WF (1994) Computing Genetic Similarity Coefficients from RAPD Data: correcting for the effects of PCR artifacts caused by variation in experimental conditions. PCR methods and Applications 4:38–43.

31. Ramos JR, M.P.C. Telles, J.A.F. Diniz-Filho, T.N. Soares, D.B. Melo and G. Oliveira (2008).Optimizing reproducibility evaluation for random amplified polymorphic DNA markers Genetics and Molecular Research 7: 1384–91

32. Russel SJ and Norving P (2004). Inteligência Artificial. Elsevier, Rio de Janeiro

33. Possingham H, Ball I and Andelman S (2000). Mathematical Methods for Identifying Representative Reserve Networks. In: Quantitative Methods for Conservation Biology (Ferson S and Burgman M, eds.). Springer-Verlag, New York, 291–306.

34. Holsinger KE and Lewis PO (2003). HICKORY v. 1.0. Department of Ecology & Evolutionary Biology, University of Connecticut, Storrs. Available at (http://www.eeb.uconn.edu/).

35. Yeh FC and Boyle TJB (1997). Population genetic analysis of co-dominant and dominant markers and quantitative traits. *Belg. J. Bot.* 129: 157. Popgene version 1.32. Available at (http://www.ualberta.ca/~fyeh/download.htm). Accessed March 2007

36. Rohlf, F.J.: NTSYS-PC: Numerical Taxonomy and Multivariate Analysis System. Version 2.11 T. - Exeter Software,Setauket 2000

37. Perrier,.Jacquemoud-Collet,J.P.(2006). DARwin software http://darwin.cirad.fr/darwin http://www.powermarker.net

Chapter 7

Amplified Fragment Length Polymorphism: An Invaluable Fingerprinting Technique for Genomic, Transcriptomic, and Epigenetic Studies

Ovidiu Paun and Peter Schönswetter

Abstract

Amplified fragment length polymorphism (AFLP) is a PCR-based technique that uses selective amplification of a subset of digested DNA fragments to generate and compare unique fingerprints for genomes of interest. The power of this method relies mainly in that it does not require prior information regarding the targeted genome, as well as in its high reproducibility and sensitivity for detecting polymorphism at the level of DNA sequence. Widely used for plant and microbial studies, AFLP is employed for a variety of applications, such as to assess genetic diversity within species or among closely related species, to infer population-level phylogenies and biogeographic patterns, to generate genetic maps, and to determine relatedness among cultivars. Variations of standard AFLP methodology have been also developed for targeting additional levels of diversity, such as transcriptomic variation and DNA methylation polymorphism.

Key words: AFLP, cDNA, DNA methylation, Epigenetics, Genetic diversity, Isoschizomers, Ligation, MSAP, Restriction enzymes, Transcriptomics

1. Introduction

Amplified fragment length polymorphism (AFLP) is a PCR-based fingerprinting technique that was first described by Vos et al. (1). Since then several modified protocols have been reported, but all typically include five main steps: (a) restriction of genomic DNA (see Note 1) and ligation of adaptors (most often performed together) to restricted fragments; (b) preselective PCR amplification of a subset of the restricted fragments; (c) selective PCR amplification, reducing further fragment number; (d) electrophoretic separation of amplified DNA fragments; (e) scoring and interpretation of the data. We detail below one of the protocols that uses a RedTaq polymerase.

Nikolaus J. Sucher et al. (eds.), *Plant DNA Fingerprinting and Barcoding: Methods and Protocols*, Methods in Molecular Biology, vol. 862, DOI 10.1007/978-1-61779-609-8_7, © Springer Science+Business Media, LLC 2012

The power of AFLP analysis derives from its ability to quickly generate large numbers of marker fragments for any organism, without prior knowledge of genomic sequence. In addition, AFLP requires only small amounts of starting template and, in comparison with other fingerprinting techniques such as RAPD (random amplified polymorphic DNA) and ISSR (intersimple sequence repeats) it exhibits much higher reproducibility (but see Note 2). Despite the fact that AFLP is a relatively labor-intensive method, it can be easily multiplexed and frequently used to amplify in the same batch hundreds of genomic fragments from hundreds of individuals.

When setting up an AFLP assay, an optimization step may be necessary to identify those primer combinations that will generate sufficient polymorphic marker fragments for a study (see Note 3). The researchers will usually choose the best three primer combinations from a dozen that have been tested across few individuals, enabling them to identify the optimal pairs for a given organism without having to design, synthesize, or perform quality control tests of their primers. The success of an AFLP assay depends on four factors: optimized reagents (see Subheading 2), standardized reaction conditions (see Note 4), a robust and reliable electrophoresis platform (see Note 5), and accurate sizing software.

While the standard AFLP protocol normally starts from genomic DNA to target genome-wide genetic loci, modifications of it have been developed to target other levels of information, such as the transcriptome and DNA methylation variation. Although expressed DNA segments (coding genes together with promoter regions and other regulatory elements) make up only a limited fraction of the genome, they are exactly the elements producing differences between phenotypes. A comprehensive transcript-profiling method that has been used to compare gene expression between individuals and groups of phenotypes is the cDNA-AFLP (2–6). For this purpose, the standard AFLP procedure (1) is performed on cDNA (the complement of mRNA) generated from a given tissue (see Note 6). Compared with other transcriptomic techniques, cDNA-AFLP is fast, requires only a small quantity of mRNA, can be applied in the absence of prior sequence data, is reproducible and sensitive, and exhibits a good correlation with Northern analysis or RT-PCR (2, 4).

Another AFLP-derived protocol is methylation-sensitive amplified polymorphism (MSAP) that targets the variation in the distribution of DNA methylation in given tissues/cell lines (7–10). DNA methylation signals are chemical modifications of the DNA bases (especially cytosines) that are relatively stable over cell generations but can differ dramatically between individuals or even between cell types within one individual. They are essential for regulating development through their influence on gene transcription; they are also involved in a number of specific biological processes, such as gene silencing and mobile element control (6). In the MSAP protocol (see Note 7), the standard AFLP technique is applied to genomic DNA by making use of two isoschizomers (methylation-sensitive/insensitive restriction enzymes that recognize the same DNA sequence, e.g., *MspI* and *HpaII*) as frequent cutters in parallel restriction reactions. A difference in banding patterns indicates methylation variation. Most of the polymorphic MSAP fragments are probably made up of noncoding regions; nevertheless, a fraction of MSAP variation among individuals is expected to correspond to epigenetic gene silencing/activation.

2. Materials

Most chemicals to be used in an AFLP assay can be bought ready-made. As much as possible the same chemicals (i.e., vial or batch) and in the same concentration should be used for all individuals to be analyzed in the AFLP experiment. Use always ultrapure water (hereafter ddH$_2$O), prepared by purifying deionized water to attain a sensitivity of 18 MΩ at 25°C or buy, e.g., Purified Water (Oxoid Ltd, UK). Store all reagents at –20°C (unless indicated otherwise).

2.1. DNA Template Preparation

1× TE buffer: dissolve 10 mM Tris–HCl (or use 10 mM of Tris, and adjust to pH 8 with HCl) and 1 mM EDTA (ethylenediaminetetraacetic acid) in 1 l ddH$_2$O, and adjust to pH 8. Store at room temperature.

2.2. Restriction-Ligation

1. MseI restriction endonuclease (the "frequent cutter"—recognizes a four-base motif, i.e., 5′-TTAA). Per individual to be analyzed 1 U MseI will be needed. Restriction enzymes are thermo-sensitive and should be taken as briefly as possible out of the freezer. Usually delivered in a solution containing glycerol, therefore, it will not freeze at –20°C.

2. EcoRI restriction endonuclease (the "rare cutter"—recognizes a six-base motif, i.e., 5′-GAATTC). Per individual to be analyzed 5 U MseI will be needed. Restriction enzymes are thermo-sensitive and should be taken as briefly as possible out of the freezer. Usually delivered in a solution containing glycerol, therefore, it will not freeze at –20°C.

3. MseI-adaptor pair: 5′-GACGATGAGTCCTGAG and 5′-TAC TCAGGACTCAT. Store each adaptor primer individually at –20°C as stock with concentration of 100 μM. Immediately prior to adding to the restriction-ligation (RL) reaction mix in proportion 1:1 (to obtain a concentration of 50 μM for each), then denature (i.e., heat up at 95°C for 5 min) the required amount of combined MseI adaptors and allow slow renature

(let them cool slowly at room temperature for ca. 10 min) to form double-stranded adaptor. Spin briefly.

4. EcoRI-adaptor pair: 5′-CTCGTAGACTGCGTACC and 5′-AATTGGTACGCAGTCTAC. Store each adaptor primer individually at –20°C as stock with concentration of 100 μM. Immediately prior to adding to the RL reaction mix in proportion 1:1:18 with ddH$_2$O (to obtain a concentration of 5 μM for each adaptor), then denature the required amount of combined EcoRI adaptors and allow slow renature. Spin briefly.

5. T4 DNA ligase. Per individual to be analyzed 0.6 U T4 DNA ligase will be needed. Restriction enzymes are thermo-sensitive and should be taken as briefly as possible out of the freezer. Minimize the time and intensity of any spin or vortex of the ligase vial. Usually delivered in a solution containing glycerol, therefore, it will not freeze at –20°C.

6. T4 DNA ligase buffer. Usually delivered with the DNA ligase. Contains 50 mM Tris–HCl, 10 mM MgCl$_2$, 1 mM ATP, and 10 mM dithiothreitol in a solution of 7.5 pH at room temperature. Can precipitate with time; before use it should be placed at 37°C and be vortexed thoroughly from time to time until fully homogenized. It is sensitive.

7. BSA (bovine serum albumin). Store as a stock solution of 10 mg/ml. Dilute prior to use to a working concentration of 1 mg/ml.

8. 0.5 M NaCl.

9. 1× TE 0.1 M buffer: dissolve 20 mM Tris–HCl (or use 20 mM of Tris and adjust to pH 8 with HCl) and 0.1 mM EDTA in 1 l ddH$_2$O, and adjust to pH 8. Store at room temperature.

10. 10× TBE buffer (stock solution): dissolve 108 g Tris base, 55 g boric acid, and 8.1 g Na$_2$EDTA in 1 l ddH$_2$O. Correct acidity to pH 8.2–8.3. Store at room temperature. The concentration of the working solution is 1× TBE buffer.

11. Size ladder with fragments in the range smaller than 1,500 bp.

12. Electrophoresis loading buffer: buy a ready-made buffer or prepare yourself one, e.g., Orange G Loading buffer. Dissolve 1.5 g Ficoll 400 and 0.015 g Orange G in 1 ml 0.5 M EDTA, and make volume up to 10 ml with ddH$_2$O.

2.3. Preselective PCR Amplification

1. Taq DNA polymerase, e.g., RedTaq (Sigma-Aldrich) or AmpliTaq (Life Technologies). DNA polymerases are thermo-sensitive and should be taken as briefly as possible out of the freezer. Minimize the time and intensity of any spin or vortex of the Taq vial. The protocol given in Subheading 3 uses RedTaq that contains an inert red dye. The dye enables visual recognition of reactions to which polymerase has been added,

confirmation of complete mixing, as well as direct loading of PCR products onto an agarose gel without the addition of electrophoresis loading buffer.

2. Taq DNA polymerase buffer. Use the recommended buffer for the chosen Taq polymerase. The protocol given in Subheading 3 uses 10× RedTaq PCR Reaction buffer (Sigma-Aldrich Co.). Some amounts of buffer are usually delivered with the DNA polymerase, but additional quantities may have to be ordered separately.

3. Deoxynucleotide Mix (dNTPs) in concentration of 10 mM each dATP, dCTP, dGTP, dTTP. We recommend the use of a ready-made mix (e.g., GeneAmp dNTP Blend, 10 mM from Life Technologies). If the dNTPs are purchased individually, a mixture has to be performed to achieve the required concentration. Aliquots of the dNTPs mixture can be stored at –20°C for several months.

4. EcoRI primer: 5′-GACTGCGTACCAATTCA (see Note 8). Store as stock solution of 100 μM.

5. MseI primer: 5′-GATGAGTCCTGAGTAAC (see Note 8). Store as stock solution of 100 μM.

6. 1× TE 0.1 M buffer (prepared as above).

7. Size ladder with fragments in the range smaller than 1,000 bp.

2.4. Selective PCR Amplification

1. RedTaq (see Subheading 2.3).

2. RedTaq buffer (see Subheading 2.3).

3. dNTPs (see Subheading 2.3)

4. EcoRI primers: 5- GACTGCGTACCAATTCXXX (see Note 8) where X stands for selective nucleotides (they vary after case). The EcoRI primers are usually fluorescently labeled. The working concentration of the EcoRI selective primer is 1 μM. As fluorescently labeled primers, the EcoRI selective primers are light-sensitive. Store as stock solution (100 μM) for several years and as working solution (1 μM) for several months.

5. MseI primers: 5-GATGAGTCCTGAGTAAXXX (see Note 8) where X stands for selective nucleotides (they vary after case). The working concentration of the MseI selective primer is 5 μM. Store as stock solution (100 μM) for several years and as working solution (5 μM) for several months.

2.5. Separation and Visualization of Fragments

1. Sephadex G-50 Fine or Superfine (Sigma Aldrich). Weigh 10 g of powder and mix with 120 ml ddH$_2$O and 100 μl 100× TE buffer. Let it stand for a couple of hours. Store at room temperature and use within 1 week. The solution of Sephadex settles out, you must resuspend it before using.

2. Multi-Screen HV plates (Millipore). Store at room temperature.

3. GeneScan ROX (Life Technologies) or another fluorescently labeled, internal ladder suitable for sequencers. Store at 4°C.

4. Hi-Di formamide (Life Technologies).

5. Polymer and buffers, specific for the type of sequencer used. Usually stored at 4°C.

3. Methods

Carry out all steps on ice, unless otherwise specified. Allow frozen chemicals (except enzymes) to fully thaw and vortex the vials briefly before using.

3.1. DNA Template Preparation

1. The AFLP procedure (1) requires c. 500 ng of genomic DNA (see Note 9) in 5.5 μl volume as starting material. Therefore, proceed to quantify DNA. To obtain the desired concentration the DNA extract can be concentrated in a vacuum oven at 60°C or diluted using 1× TE buffer. This entire step can be performed at room temperature.

3.2. Restriction-Ligation

1. Calculate the appropriate amount of each chemical for the required number of reactions according to Table 1. For the three enzymes calculate the required amount according to their concentration (see Note 10). Do not forget to account also for replicates, blanks (see Note 2) and two more samples as a tolerance for potential pipeting inaccuracies.

Table 1
Chemicals necessary for RL

Chemicals	μl per sample	Comments
10× T4 ligase buffer	1.1	Vortex before use
0.5 M NaCl	1.1	
1 mg/ml BSA	0.55	Freshly diluted
MseI adaptor pair (50 μM)	1	Already annealed
EcoRI adaptor pair (5 μM)	1	Already annealed
MseI restriction endonuclease	… (1 unit)	
EcoRI restriction endonuclease	… (5 units)	
T4 DNA ligase	… (0.6 unit)	
ddH$_2$O (to make the total volume of 5.5 μl per sample)	……	

2. Heat the required amount of MseI and EcoRI adaptor pairs at 95°C for 5 min, each pair in a separate vial. Allow them to cool gradually to room temperature for c. 10 min. Spin briefly in a microcentrifuge for 10 s.

3. Prepare a master mix for all samples that you plan to analyze in one batch, starting with ddH$_2$O, T4 ligase buffer, NaCl, BSA, both adaptor pairs and finishing with the three enzymes (see Note 11). Spin briefly.

4. Aliquot 5.5 µl of the master mix in individual tubes.

5. For each sample, add 5.5 µl DNA (containing 500 ng—see Subheading 3.1) in one tube. The final reaction volume will be 11 µl. Vortex and centrifuge briefly.

6. Incubate the reactions at 37°C for at least 3 h in a thermal cycler with heated cover. Continue to incubate at 17°C overnight, or at least for 3 h (17°C is the optimum temperature for ligation activity). In case of PCR machines without heated lid, the reaction mix has to be covered with a drop of mineral oil; evaporation leads to EcoRI Star Activity (nonsite-specific cutting; see Note 10).

7. You can test the efficiency of the restriction reaction (see Notes 12 and 13) by running 5 µl of several of the reactions (mixed with loading buffer) on a 1.5% agarose gel in 1× TBE buffer (see Note 14) for 20 min at 90 V. When preparing the agarose gel add ethidium bromide (attention! carcinogenic) or GelRed to view the RL product under UV lamp (Fig. 1d–f).

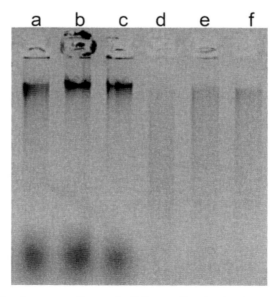

Fig. 1. Examples of agarose gel images for DNA extractions (**a–c**) and successful RL (**d–f**). see also Notes 12 and 13.

Table 2
Chemicals necessary for the preselective and selective amplification

Chemicals	µl per preselective sample	µl per selective sample
ddH$_2$O	5.86	5.50
10× RedTaq Buffer	1.14	1
Primers	0.58 of the mixed primer pair	0.54 MseI primer (5 µM) 0.54 EcoRI primer (1 µM)
dNTPs (10 mM)	0.22	0.22
RedTaq (1 U/µl)	0.2	0.2

8. Stop the reaction by diluting it 20-fold with 1× TE 0.1 M buffer. Mix thoroughly. For the samples for which an aliquot of the PCR product has been run on agarose gel (in step 7), do not forget to reduce the dilution volume.

9. Store the diluted RL reactions in the fridge for 1 day. The RL reactions can be stored for months at –20°C.

3.3. Preselective PCR Amplification

1. Calculate the appropriate amount of each chemical for the required number of reactions according to Table 2. Carry over all replicates and blanks. Do not forget to account also for two more samples as a tolerance for potential pipeting inaccuracies.

2. Starting from the stock solutions (100 µM), dilute and mix preselective primers in proportion of 1:1:18 with ddH$_2$O to result in a working concentration of 5 µM each primer.

3. Prepare a master mix for all samples that you plan to analyze in one batch, starting with ddH$_2$O, 10× Taq-Buffer, dNTPs, primers and finishing with the Taq (see Note 11). Spin briefly.

4. Aliquot 8 µl of the master mix in individual tubes.

5. Add 2 µl of the diluted RL product to each tube. The final reaction volume will be 10 µl. Vortex and centrifuge (500×g) briefly.

6. Use a thermal cycler with heated cover and run the following program: one hold of 72°C for 2 min, 20 cycles of (94°C for 1 s, 56°C for 30 s and 72°C for 2 min), and finish with a hold of 60°C for 30 min. Program the cycler to keep the reactions at 4°C until you remove them.

7. You can test the efficiency of the preselective amplification by running 5 µl of several of the reactions on a 1.5% agarose gel in 1× TBE buffer (see Note 14) for 20 min at 90 V. If using the RedTaq polymerase no loading buffer has to be used.

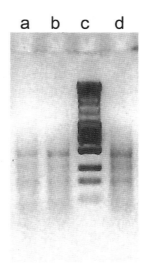

Fig. 2. Examples of successful preselective reactions (**a, b, d**). (**c**) Ladder.

When preparing the agarose gel add ethidium bromide (attention! carcinogenic) to view the RL product under UV lamp (Fig. 2). A smear product with few brighter bands in the 100–1,500 base pair range should be visible.

8. Diluting the preselective reactions 20-fold with 1× TE 0.1 M buffer. Mix thoroughly. For the samples for which an aliquot of the PCR product has been run on agarose gel (in step 7), do not forget to reduce the dilution volume.

9. Store the diluted preselective reactions in the fridge for 1 day and at –20°C for months.

3.4. Selective PCR Amplification

1. Calculate the appropriate amount of each chemical for the required number of reactions according to Table 2. Carry over all replicates and blanks. Do not forget to account also for two more samples as a tolerance for potential pipeting inaccuracies.

2. Prepare a master mix for all samples that you plan to analyze in one batch, starting with ddH_2O, 10× Taq-Buffer, dNTPs, primers and finishing with the Taq (see Note 11). Spin briefly. If you are doing several primer combinations, the master mix for each combination must be prepared in separate vials.

3. Aliquot 8 μl of the master mix in individual tubes. If you are doing several primer combinations, aliquot each combination in separate tubes.

4. Add 2 μl of the diluted preselective product to each tube. The final reaction volume will be 10 μl. Vortex and centrifuge ($500 \times g$) briefly.

5. Use a thermal cycler with heated cover and run the following program (90% ramp time): one hold of 94°C for 2 min, 9 cycles of (94°C for 1 s, 65°C—1°C every cycle for 30 s and 72°C for 2 min), followed by 23 cycles of (94°C for 1 s, 56°C for 30 s and 72°C for 2 min) and finish with a hold of 60°C for 30 min. Program the cycler to keep the reactions at 4°C until you remove them.

6. We recommend freezing the selective reactions as soon as possible. They can, however, be kept for 1 day in the fridge. If necessary, selective reactions can be stored for several months at –20°C, but we recommend running them as soon as possible in the sequencer.

3.5. Separation and Visualization of Fragments

1. Apply 200 µl of mixed Sephadex solution to each well of a Multi-Screen (MS) plate. Place the MS plate on top of a microtiter plate to collect water. Pack the Sephadex by spinning at $600 \times g$ for 1 min. Discard water that has been collected in the microtiter plate.

2. Repeat step 1.

3. Repeat once again step 1, but this time pack the Sephadex by spinning at $600 \times g$ for 5 min.

4. Place MS plate with the Sephadex filter on top of a clean microtiter plate to collect the filtered selective product (see Note 15).

5. Mix together the selective reactions of up to three primer combinations corresponding to one individual sample, by applying 5 µl each selective PCR product (e.g., labeled green, yellow and blue). Spin the MS plate (on top of the clean microtiter plate) at $600 \times g$ for 5 min (see Note 16).

6. Discard the Sephadex filter. Wash the HV plate with ddH$_2$O and let it dry at room temperature. The HV plate can be reused for up to ten times.

7. Make up the loading mixture for the number of samples to be loaded on the sequencer using 9.8 µl Hi-Di formamide and 0.2 µl of GeneScan ROX per sample. Do not forget to account also for two more samples as a tolerance for potential pipeting inaccuracies.

8. Aliquot 10 µl of loading mixture to each well of a clean microtiter plate.

9. Add 1.2 µl of the filtered, combined selective products to each well. Vortex and centrifuge briefly.

10. Cover the microtiter plate containing loading mixture and sample; heat it up at 95°C for 5 min. Cool the plate on ice immediately to denature the AFLP fragments.

11. Load the plate containing the denatured samples onto the sequencer.

4. Notes

1. The AFLP technique requires good quality DNA; degraded DNA will produce noncomparable profiles between individuals. The best is to use silica gel dried material.

2. In standardized conditions, AFLP profiles usually display high levels of reproducibility. However, reproducibility has to be always tested for each batch of reactions by using replicates. It has been recommended to replicate c. 20% of the samples (11). They will then be used to calculate the error rate as the number of errors divided by the number of phenotypic comparisons within replicated samples. Acceptable error rates for AFLP are lower than 5% (11). In addition to replicates, always use negative controls to test for possible systematic contamination. This is very important as any DNA present in the reaction will produce fragments, which cannot be distinguished from the targeted products. In most cases, negative controls will produce some peaks in the low base pair range due to the amplification of excess preselective primers in the selective amplification (1).

3. We recommend choosing six individuals that may be representative for the samples to be analyzed (e.g., select individuals from different regions, or different taxa). Look for primer combinations that produce an appropriate number of fragments (c. 100), that are evenly distributed across the range to be analyzed (50–500 bp). By combining different restriction enzymes (e.g., by using frequently cutting enzymes only) and using primers with two to four selective nucleotides, the number of AFLP fragments per profile can be optimized. see also Note 8.

4. Take measures to standardize each of the AFLP steps. Run all samples in a minimum amount of time. Use as much as possible the same PCR machine (or at least the same model) for all samples investigated for each primer pair. This will contribute towards keeping the error rate to a minimum (see Note 1).

5. Always use the same internal size standard for each sample run on the sequencer. Make sure to perform the sizing of the AFLP fragments based on the same fragments from the internal size standard.

6. Extract total RNA (e.g., with SV Total RNA Isolation System, Promega) from tissue material fixed in RNAlater (Sigma) and stored at $-20°C$. Then, synthesize cDNA from mRNA (e.g., with SuperScript™ Double-Stranded cDNA Synthesis Kit, Invitrogen), using an oligo $(dT)_{12-18}$ primer. Apply the standard AFLP protocol further to obtain cDNA-AFLP fingerprints. As cDNA comprises much less DNA than the entire genome, use selective primers with only two selective nucleotides (or one primer with two selective nucleotides and one with three).

7. The MSAP technique uses two methylation-sensitive isoschizomers (e.g., MspI and HpaII) as frequent cutters, each in combination with the same rare cutter (e.g., EcoRI) in parallel reactions (7). The two isoschizomers recognize the same sequence (5'-CCGG) but differ in their sensitivity to DNA methylation. Comparison of the two profiles for each individual allows assessment of the methylation state of the restriction sites. ^{Me}CpG sites are recognized by MspI only, whereas plant-specific ^{hemiMe}CpCpG sites are recognized by HpaII only (9, 10). Sites that are hypermethylated (i.e., both at the internal and external Cs), and sites that are fully methylated at the external Cs (i.e., on both strands) are not cut by either enzyme, whereas sites that are free from methylation are recognized by both. Normally the MSAP protocol uses an excess of restriction enzymes (10 U per sample of each restriction enzyme—use highly concentrated enzymes) in the RL, which uses also longer times (overnight).

8. According to the genome size and the abundance of restriction sites in the genomes to be analyzed you may need to increase or decrease the number of selective nucleotides of the primers. For a small genome use shorter selective primers, with only two selective nucleotides. For a large genome use longer pre-selective (with two selective nucleotides) and selective (with four selective nucleotides) primers. The tuning of the number of AFLP fragments can also be obtained by using a different combination of restriction enzymes (e.g., two "rare" cutters or two "frequent" cutters).

9. Genomic DNA should be extracted from similar amounts of tissue (c. 10 mg) to ensure a comparable concentration of DNA extractions across all samples.

10. Nonstandard conditions that can induce Star Activity of EcoRI include high glycerol concentration, an excessive amount of enzyme used, low salt concentration, high pH (pH > 8), and contamination with organic solvents (see http://www.neb.com/nebecomm/products/faqproductR0101.asp#1). To prevent Star Activity avoid evaporation, use highly concentrated enzymes (to reduce the amount of glycerol introduced in the reaction) and use as little amounts of enzyme as possible to ensure complete digestion.

11. Always start preparing Master Mixes with the largest volumes (ddH$_2$O and/or buffers) and add any enzymes the last.

12. A smear product in the 100–1,500 base pair range should be visible. Make sure the genomic DNA is fully restricted, so no high-weight DNA molecules are present (Fig. 1d–f).

13. The efficiency of the ligation step cannot be verified on the gel.

14. Prepare the agarose gels by boiling for 2 min 0.75 g agarose in 50 ml TBE buffer. Let the solution to cool to 70°C and add 1 μl ethidium bromide (attention! Highly carcinogenic).

15. Sephadex purification of the AFLP products leads to a reduction of primer dimers and a better resolution in the range 50–500 bp. In the protocol presented, the purification step is necessary to remove from the selective products the inert red dye that has been introduced with the RedTaq polymerase. Other purification protocols can also be used.

16. Packing of the Sephadex filter and filtering of the PCR products must be performed at the same rotation speed.

Acknowledgment

O.P. was financed by an Austrian Science Fund (FWF) project (P222260-B16).

References

1. Vos P, Hogers R, Bleeker M et al (1995) AFLP: a new technique for DNA fingerprinting. *Nucleic Acids Research* 23:4407–4414.

2. Bachem CWB, van der Hoeven RS, de Bruijn SM et al. (1996) Visualization of differential gene expression using a novel method of RNA fingerprinting based on AFLP: analysis of gene expression during potato tuber development. *The Plant Journal* 9:745–753.

3. Kuhn E (2001) From library screening to microarray technology: strategies to determine gene expression profiles and to identify differentially regulated genes in plants. *Ann. Bot.-London* 87:139–155.

4. Donson J, Fang Y, Espiritu-Santo G et al. (2002) Comprehensive gene expression analysis by transcript profiling. *Plant Molecular Biology* 48:75–97.

5. Breyne P, Dreesen R, Cannoot B et al. (2003) Quantitative cDNA-AFLP analysis for genome-wide expression studies. *Molecular Genetics and Genomics* 269:173–179.

6. Paun O, Fay MF, Soltis DE, Chase MW (2007) Genetic and epigenetic alterations after hybridization and genome doubling. *Taxon* 56:649–656.

7. Baurens F-C, Bonnot F, Bienvenu D et al. (2003) Using SD-AFLP and MSAP to assess CCGG methylation in the banana genome. *Plant Molecular Biology Reporter* 21:339–348.

8. Ainouche ML, Baumel A, Salmon A, Yannic G (2003) Hybridization, polyploidy and speciation in *Spartina* (Poaceae). *New Phytologist* 161:165–172.

9. Salmon A, Ainouche ML, Wendel JF (2005) Genetic and epigenetic consequences of recent hybridization and polyploidy in *Spartina* (Poaceae). *Molecular Ecology* 14:1163–1175.

10. Paun O, Bateman RM, Fay MF et al. (2010) Stable epigenetic effects impact adaptation in allopolyploid orchids (*Dactylorhiza*: Orchidaceae). *Molecular Biology and Evolution* 27:2465–2473.

11. Bonin A, Bellemain E, Bronken Eidesen P et al. (2004) How to track and assess genotyping errors in population genetic studies. *Molecular Ecology* 13:3261–3273.

Chapter 8

Restriction Fragment Length Polymorphism of the 5S-rRNA-NTS Region: A Rapid and Precise Method for Plant Identification

Cinzia Margherita Bertea and Giorgio Gnavi

Abstract

Molecular genetic methods have several advantages over classical morphological and chemical analyses. The genetic method requires genotype instead than phenotype, therefore PCR-based techniques have been widely used for a rapid identification of plant species, varieties and chemotypes. Recently, the molecular discrimination of some higher plant species has been evaluated using sequences of a 5S-rRNA gene spacer region. The variation in the nontranscribed sequence (NTS) region has been used in a number of plant species for studying intraspecific variation, genome evolution, and phylogenetic reconstruction. Here, we describe a rapid method based on the use of the 5S-rRNA-NTS region as a tool for plant DNA fingerprinting, which combines PCR, sequencing and restriction fragment length polymorphism analyses.

Key words: DNA fingerprinting, 5S-rRNA gene, Nontranscribed sequence region, PCR–RFLP, High-resolution gel capillary electrophoresis, Agilent 2100 Bioanalyzer

1. Introduction

Morphological, anatomical, and chemical analyses aimed to detect and quantify plant samples are affected by environmental and/or developmental factors or by method of sample storage (1). This means that, according to environmental conditions, the same genotype may express different chemical patterns, or, conversely, different genotypes may respond to the same environmental pressure with the same phenotypic expression (2). Very often, the identification of plant samples in a mixture is difficult to achieve and this problem is particularly exacerbated when plant mixtures are powdered. In this context, molecular genetic methods have recently been shown to be very effective in genotypic discrimination. In the last few decades, DNA-based experiments have become widely employed

Nikolaus J. Sucher et al. (eds.), *Plant DNA Fingerprinting and Barcoding: Methods and Protocols*, Methods in Molecular Biology, vol. 862, DOI 10.1007/978-1-61779-609-8_8, © Springer Science+Business Media, LLC 2012

techniques for a rapid identification of plant material. By using PCR approaches, nanogram quantities of DNA are required to amplify and yield sufficient amounts of template DNA for molecular genetic analysis (3).

Recently, the phylogenetic relationship of some higher plant species has been evaluated using sequences of a 5S-rRNA gene spacer region. The 5S-rRNA genes are present as multiple copies arranged in tandem arrays in the nuclear genome of higher plants (4).

Each 5S-rRNA repeat unit consists of a 120-bp coding sequence and a nontranscribed spacer (NTS). Although the transcribed region is conserved in plant species, the NTS commonly shows length heterogeneity (in the range of 100–800 bp) and extensive sequence divergence (4–6). Sequence conservation of coding regions and high divergence in the spacer regions provided a good model for studying the organization and evolution of multigenes in different plant species (7, 8). Based on these assumptions, the variation in the NTS region has been used in a number of plant species for studying intraspecific variation, genome evolution, and phylogenetic reconstruction (9).

Since the introduction of PCR in 1986, an ever-increasing number of scientific applications have been reported, including the direct cloning, mutagenesis, sequencing and exact engineering of specific genes or gene sequences directly from genomic DNA or complementary DNA (cDNA) (10). Detection of the PCR products is performed by agarose gel electrophoresis, ethidium bromide staining, ultraviolet irradiation and comparison of the product size with a DNA size marker. New methods of PCR product detection, such as the application of high-resolution capillary gel electrophoresis (e.g., Agilent 2100 Bioanalyzer), have greatly improved the accuracy of PCR product visualization and have opened up new possibilities in applying this technique for plant identification (2, 11, 12). Sequence polymorphisms in PCR amplification products can be detected using restriction digestion (2, 11–13). The bands on gels, which typically serve as molecular markers, may arise from cutting DNA at specific sites with restriction enzymes, in order to detect restriction fragment length polymorphisms (RFLPs).

Molecular fingerprinting using the 5S-rRNA-NTS region allowed the rapid and precise identification of several aromatic plants (14). By aligning the isolated nucleotide sequences, great diversities can be found in the spacer regions, this allows the design of specific primers. In addition, a PCR–restriction fragment length polymorphism (PCR–RFLP) method can be applied using restriction enzymes (2, 11, 12). Thus, molecular analysis of the 5S-rRNA-NTS can lead to the precise and unequivocal identification of a species (14). The use of the Agilent 2100 Bioanalyzer and lab-on-a-chip technology allows for rapid separation of dsDNA and delivers higher quality data than classical slab-gel electrophoresis, in terms of band separation and resolution (Fig. 1).

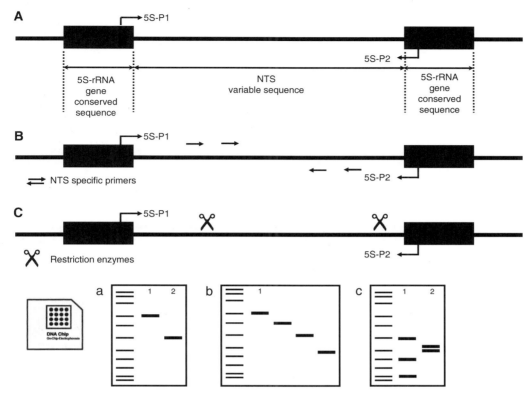

Fig. 1. Outline of the main steps of the 5S-rRNA-NTS PCR–RFLP method employed for identification of plant species, varieties, chemotypes, cytotypes, etc. The first PCR reaction is carried out by using 5S-rRNA-NTS primers designed on the conserved coding regions of the 5S-rRNA gene (**A**). The resulting bands are then purified and subcloned into a plasmid vector before sequencing (see Subheadings 3.1 and 3.2). Sequence analyses allow to design specific primers on the nontranscribed spacer (NTS) region to be employed in PCR reactions (**B**) and to identify restriction enzyme sites for restriction fragment length polymorphism (RFLP) analyses (**C**). The products of the steps **A**, **B**, and **C** are visualized by using DNA 1000 LabChip® Kit and the Agilent 2100 Bioanalyzer. This capillary gel electrophoresis technique provides higher resolution of smaller fragments in a short time (about 30 min). (**a**) PCR products generated by primers 5S-P1 and 5S-P2 on plant species 1 and 2; (**b**) PCR products generated by primers 5S-P1 and 5S-P2 in combination with specific NTS primers on the plant species 1; (**c**) RFLP analyses on plant species 1 and 2.

2. Materials

Prepare all solutions, not included in kits, using ultrapure water (prepared by purifying de-ionized water to attain a sensitivity of 18 MΩ cm at 25°C) and analytical grade reagents. Prepare and store all reagents at room temperature (unless indicated otherwise). Diligently follow all waste disposal regulations when disposing waste materials.

2.1. PCR Amplification and Agarose Gel Electrophoresis Components

1. Ultrapure water for PCR reaction. Store at room temperature.

2. 10× PCR Buffer, *Taq* DNA polymerase, 10 mM dNTP mix (Fermentas International Inc., Burlington, Canada). Store at –20°C.

3. Primers (Invitrogen BV, Leek, The Netherlands): lyophilized primers are dissolved in ultrapure water. Add 100 µL of ultrapure water each 100 nmol of lyophilized primers to get a 100-µM stock solution. Make 10 µM working solution by diluting 1:10 an aliquot of the stock solution. Store both solutions at −20°C.

4. 0.2 mL PCR tubes.

5. Tris-acetate–EDTA (TAE) Buffer: 40 mM Tris-acetate, 1 mM EDTA (pH 8.5).

6. Make a concentrated (50×) stock solution of TAE by weighing out 242 g Tris base (FW = 121.14) and dissolving in approximately 750 mL ultrapure water in a graduated cylinder or in a glass beaker (see Note 1). Carefully add 57.1 mL of glacial acid and 100 mL of 0.5 M EDTA, pH 8.0 (see Note 2). Adjust the pH to 8.5 with 5N NaOH. Make up to 1 L with water. Store at room temperature. The working solution of 1× TAE buffer is made by simply diluting the stock solution by 50× in ultrapure water.

7. Loading buffer (6×): 0.25% bromophenol blue, 30% glycerol. Weigh 25 mg bromophenol blue. Add 3 mL glycerol and make up volume to 10 mL with sterile water. Store at 4°C.

8. Ethidium bromide solution: 10 mg/mL ethidium bromide in ultrapure sterile water. Weigh 100 mg of ethidium bromide (see Note 3) and dissolve it in 10 mL of ultrapure sterile water. Store at 4°C in the dark with toxic labels on it.

2.2. Ligation of the PCR products, Cloning, and Transformation Components

1. pGEM®-T Easy Vector System I (Promega Corporation, Madison, WI, USA).

2. Subcloning™ DH5α™ Efficiency Competent Cells (Invitrogen, Carlsbad, CA, USA).

3. 15 mL plastic or glass tubes for microbiology.

4. 1.5 mL microfuge sterile tubes.

5. Luria–Bertani (LB) medium and agar plates: dissolve 10 g Bacto-tryptone, 5 g Bacto-yeast extract, 5 g NaCl in 950 mL de-ionized water. Adjust the pH to 7.0 with 5N NaOH. Adjust the volume to 1 L with de-ionized water. For LB/agar plates add 20 g/L bacteriological agar to the broth. Sterilize by autoclaving (121°C, 15 min). For ampicillin bacterial selection add 200 µL of a 100-mg/mL ampicillin solution (see Note 4) to 200-mL LB medium/LB agar (100 µg/mL final ampicillin concentration) (see Note 5). Store liquid medium at room temperature and LB/agar plates, wrapped with Parafilm M®, at 4°C.

2.3. High-Resolution Capillary Gel Electrophoresis Components

1. Agilent DNA 1000 kit (catalog number 5067-1504) (Agilent Technologies Inc., Santa Clara, CA, USA): 25 chips and reagents (see Note 6) designed for sizing and analysis of DNA fragments.

2. Chip priming station (catalog number 5065-4401) (Agilent Technologies).

3. Gel-dye mix. Prepare the gel-dye mix by allowing DNA dye concentrate (blue cap) and DNA gel matrix (red cap) to equilibrate to room temperature for 30 min. Vortex DNA dye concentrate, spin down briefly, and add 25 µL of the dye to a DNA gel matrix vial. Vortex solution well and spin down briefly. Transfer to spin filter (provided with the kit). Centrifuge at $2,240 \times g$ for 15 min. Protect solution from light, by wrapping the tube with aluminum foil. Store this solution at 4°C up to 4 weeks (see Note 7).

3. Methods

3.1. PCR Amplification, Agarose Gel Electrophoresis Analysis, and Band Purification

1. Prepare 50 µL PCR reaction by mixing 1 µL of genomic DNA (about 20 ng) (see Note 8), forward 5S-P1 (5'GTGCTT GGGCGAGAGTAGTA-3') primer and 2 µL of 10 µM (20 pmol) reverse primer 5S-P2 (5'-TTAGTGCTGG TATGATCGCA-3') designed on the flanking conserved regions of the NTS of 5S-rRNA gene (3, 8) (see Note 9), 5 µL of 10× PCR buffer, 1 µL 10 mM dNTPs (0.2 mM final concentration), 0.5 U of *Taq* DNA polymerase (Fermentas), and sterile water to 50 µL.

2. Run the PCR reaction according to the following cycling conditions: initial denaturation: 2 min at 94°C followed by 1 min denaturing at 94°C, 1 min annealing at 56°C, and 2 min elongation at 72°C repeated for 35 cycles and with 5 min extension at 72°C (see Note 10).

3. Prepare the 2% agarose gel (see Note 11). Weigh out 2 g of agarose into a 250-mL Erlenmayer flask. Add 100 mL of 1× TAE, swirl to mix.

4. Warming in a microwave oven for about 1 min to dissolve the agarose (see Note 12). Leave it to cool on the bench for 5 min down to about 60°C. Add 1 µL of ethidium bromide (10 mg/mL) and swirl to mix (see Note 13). Pour the gel slowly into the tank. Push any bubbles away to the side using a disposable tip. Insert the appropriate comb. Leave to set for at least 30 min with the lid on if possible. Pour the running buffer (1× TAE buffer) into the gel tank to submerge the gel to 2–5 mm depth.

5. Prepare the samples. Transfer 40 µL of each sample to a fresh 1.5-mL microfuge tube. Add about 7 µL of 6× loading buffer to each sample (see Note 14).

6. Load the gel with samples and use a well or more for ladder (see Note 15). Close the gel tank, switch on the power-source, and run the gel at 80 V. Monitor the progress of the gel by reference to the marker dye. Switch off and unplug the gel tank and visualize the bands on the UV transilluminator (see Note 16).

7. Cut the bands of interest by using a clean scalpel (see Note 17) and put them in clean preweighed 1.5 mL microfuge tubes.

8. Proceed with band purification by using one of the available commercial kits, i.e., GFX™ PCR DNA and Gel Band Purification Kit (Amersham Biosciences) (11), following manufacturer's instructions.

3.2. Ligation of the PCR Products, Cloning, and Transformation

1. Ligate purified bands in a plasmid vector, i.e., pGEM®-T Easy Vector System I (Promega) (11). Add 5 μL of 2× Rapid Ligation Buffer, T4 DNA Ligase, 4 μL of purified PCR products, 0.5 μL of pGEM®-T Easy Vector (50 ng/μL, 0.5 μL of T4 Ligase) (Promega) (see Note 18). Incubate the ligation mixtures overnight at 4°C (see Note 19).

2. Pipette 50 μL of competent cells (*Escherichia coli* Subcloning DH5α Efficiency Competent Cells—Invitrogen) into 1.5-mL ice-cold microfuge tubes containing 10 μL of the ligation reaction and mix well by pipetting. Place the tubes on ice for 30 min.

3. Heat shock the tubes with transformation mixtures at 42°C for exactly 90 s. Return the cells to ice for 2 min.

4. Add 940 μL LB broth (without antibiotic) to the transformation mixtures and incubate the tubes at 37°C in a circulating water bath for 1 h and 30 min (see Note 20).

5. Pipette 200 μL of the transformation mixtures onto labeled ampicillin agar plates and spread them around using a sterilized, bent glass rod spreader (see Note 21).

6. Place the plates upside down in a 37°C air incubator and incubate overnight.

7. 16–20 h later, pick at least 15 of well-isolated colonies from the plate (see Note 22) with a sterile small pipette tip and dissolve each colony in 10 μL of sterile water in 0.2 mL PCR tubes (see Note 23).

8. Run a colony PCR to check the size of the inserts ligated into the plasmids. Add 5 μL of the following PCR mix: 0.15 μL from 10 μM stock solution of forward and reverse primers (see Note 24), 1.50 μL of 10× PCR buffer (Fermentas), 0.3 μL of 10 mM dNTPs (0.2 mM final concentration), 0.3 U of *Taq* DNA polymerase (Fermentas), and sterile water to 5 μL.

9. Run the PCR reaction according to the following conditions: initial denaturation: 4 min at 94°C followed by 1 min denaturing at 94°C, 1 min annealing at 56°C, and 2 min elongation at 72°C repeated for 35 cycles and with 5 min extension at 72°C.

10. Gel electrophoresis analysis: use 5 µL of the PCR products to check quality and quantity of the amplified bands by agarose gel electrophoresis following the protocol reported above.

11. Pick up a good amount of at least ten individual positive colony streaks from the replica plate (those ones harboring inserts of the correct size) dissolve them in a 3-mL LB broth containing 100 µg/mL ampicillin in 15 mL disposable or glass tubes (see Note 25).

12. Grow 3 mL liquid bacterial cultures overnight at 37°C in a rotary shaker (250 rpm).

13. Proceed with the plasmid preparation by using one of the available commercial kits such as QIAprep Spin Miniprep Kit (Qiagen, Hilden, Germany) (11).

14. Run a PCR reaction to check the presence of inserts into the plasmids following the conditions reported above.

15. Sequence both strands of the plasmids at least twice by using vector universal primers such as SP6, T7, or M13 forward/reverse or 5S-rRNA NTS primers (5S-P1 and 5S-P2) (see Note 26).

3.3. Molecular Fingerprinting

3.3.1. PCR Amplification Using Specific NTS Primers

1. Align the sequences deriving from the sequencing of the same band and create a single consensus by using one of the available on-line alignment programs such as Clustal × 2.0. In addition, align sequences belonging to different species, varieties, or chemotypes in order to find differences at the nucleotide level.

2. Design specific primers on the consensus sequence by using one of the free internet programs, e.g., Primer3 (http://frodo.wi.mit.edu/primer3/) (see Note 27).

3. Use forward and reverse specific NTS primers in further PCR amplifications also in combination with 5S-rRNA NTS primers (5S-P1 and 5S-P2) by using the PCR conditions already described (see Note 10).

3.3.2. Restriction Length Fragment Polymorphism Analyses

1. Identify specific restriction enzyme sites for the sequences of interest by using one of the free programs available on line, e.g., CLC Sequence Viewer 6 (see Note 28).

2. Digest purified PCR products of the 5S-rRNA gene spacer region with the appropriate restriction enzyme and buffer, following manufacturer's instructions. Here, we report a typical digestion protocol: add 2 µL of 10× Digestion Buffer, 10 µL of purified PCR products, 1 µL of restriction enzyme

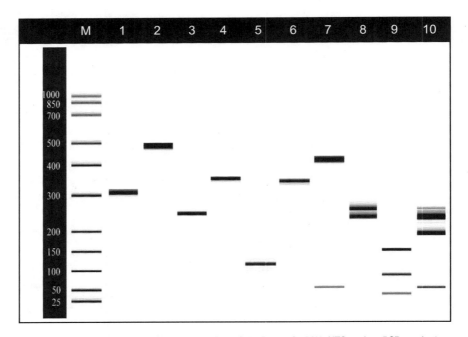

Fig. 2. An example of DNA fingerprinting by the use of PCR–RFLP of the 5S-rRNA-NTS region. PCR products generated by primers (5S-P1 and 5S-P2) flanking the spacer region of 5S-rRNA gene using *Salvia divinorum* and *Salvia officinalis* genomic DNA as template. A single fragment of approximately 300 bp was produced from *S. officinalis* (*Lane 1*) whereas a single fragment of about 500 bp was produced from *S. divinorum* (*Lane 2*). *S. divinorum* PCR products generated by the combination of 5S-rRNA-NTS primers (5S-P1 and 5S-P2) and SD primers, designed on the specific sequence of *S. divinorum*: 5S-P1 SD1, 265 bp (*Lane 3*); 5S-P1 SD2, 371 bp (*Lane 4*); SDF-SD1, 127 bp (*Lane 5*); SDF 5S-P2, 349 bp (*Lane 6*); *S. divinorum* and *S. officinalis* PCR–RFLP analysis: *S. divinorum* *Nde*I digested PCR products (428 and 59 bp) (*Lane 7*); *S. divinorum* *Taq*I PCR digested products (252 and 235 bp) (*Lane 8*); *S. officinalis* *Taq*I PCR digested products (47, 91, 161 bp. A 9 bp fragment is not visible) (*Lane 9*); *S. divinorum* *Taq*I PCR digested products subsequently digested with *Nde*I (a fragment at about 60 bp and one at about 190 bp, deriving from the cleavage of the former fragment at 252 bp, one of uncleaved product at about 235 bp, and the fourth at about 250 bp indicating an incomplete cleavage of the former 252 bp fragment) (*Lane* 10). PCR products and restriction enzyme reactions were separated using the Agilent 2100 Bioanalyzer and the DNA 1000 LabChip® Kit (Agilent Technologies) (see ref. 11).

(see Note 29), and ultrapure water up to 20 μL. Incubate for 1 h at 37°C in a water bath (see Note 30).

3. When possible, operate a second restriction analysis by cutting the fragment produced with the first restriction enzyme (see Note 31). This step provides a more precise discrimination of the material in analysis (Fig. 2).

3.3.3. High-Resolution Gel Capillary Electrophoresis for PCR Product and Restriction Enzyme Digestion Analysis

1. Analyze 1 μL of PCR products obtained with 5S-rRNA NTS general/specific primers or restriction enzyme digestion (see Note 32) by using capillary gel electrophoresis and the Agilent 2100 Bioanalyzer (Agilent Technologies).

   Allow the gel-dye mix equilibrate to room temperature for 30 min before use (see Note 33). Put a new chip from DNA 1000 LabChip® Kit on the chip priming station (see Note 34). Make sure that the chip priming station is set up properly: base plate in position C and syringe clip down to the lowest position.

Pipette 9 μL of gel-dye mix in the appropriate well as reported in the manual instruction. Make sure that the plunger is positioned at 1 mL and then close the chip priming station. Press plunger until it is held by the clip. Wait for exactly 60 s then release clip. Wait for 5 s. Slowly pull back plunger to 1 mL position. Open the chip priming station and pipette 9 μL of gel-dye mix in the remaining wells marked with gel-matrix symbol. Pipette 5 μL of marker in all 12 sample wells and ladder well. Do not leave any wells empty. Pipette 1 μL of DNA ladder in the well marked with a ladder symbol. In each of the 12 sample wells, pipette 1 μL of sample (used wells) or 1 μL of de-ionized water (unused wells) (see Note 35). Put the chip horizontally in the adapter and vortex for 1 min at the indicated setting (2,400 rpm) (see Note 36). Run the chip in the Agilent 2100 Bioanalyzer within 5 min (see Note 37), using the appropriate program design for DNA 1000 chip (Fig. 2).

4. Notes

1. Add water at the bottom of the cylinder or beaker and dissolve Tris slowly by using a magnetic stir bar. It is also possible to warm up the solution at 37°C to facilitate faster dissolving. Remember to bring back the solution at room temperature before adjusting the pH.

2. It is better to use an already prepared EDTA solution instead than dissolving EDTA powder with Tris, since the EDTA will not dissolve until the pH is about 8. To make 1 L of 0.5 M EDTA pH 8.0 solution, weigh 186.1 g of $Na_2EDTA \cdot 2H_2O$ and dissolve the powder slowly in 800 mL of ultrapure water in a cylinder. Adjust the pH to 8 with 10N NaOH solution. Make up to 1 L with water and autoclave. Store at room temperature.

3. Ethidium bromide is mutagenic and must be handled with extreme caution. Prepare ethidium bromide solution in a fume hood to prevent inhalation of dust or aerosols. Wear gloves at all times. Nitrile rubber gloves are more effective than latex ones while you are handling ethidium bromide. Dispose contaminated gloves, test tubes, paper towels, tips, etc. into a dedicated ethidium bromide waste container.

4. Ampicillin stock solution (100 mg/mL) is prepared by adding 0.5 g of ampicillin in 5 mL sterile water. After dissolving, filter sterilize the solution, since antibiotics are heat labile and they cannot be autoclaved. Use a 0.2-μm filter (Millipore). Make 200 μL aliquots and store at −20°C. Storage in aliquots helps to avoid freezing/defreezing cycles with subsequent antibiotic degradation.

5. Add the antibiotic after autoclaving when the LB broth/agar is at about 50–55°C. Check by putting the hands on the bottle. Stir or shake accurately to get a homogenous antibiotic distribution.

   For agar plates, pour about 25–30 mL in a 10-cm Petri dish. Let agar solidify at room temperature, then store plates, wrapped with Parafilm, at 4°C for later use. Stability problem is the same for medium and plates. Always add antibiotics to LB medium freshly and just before use. Ampicillin plates should be stable for 4 weeks at 4°C, but often after 1 week a reduced sensitivity is observed. After use, dispose the plates in the bio-hazard bag and autoclave them before trashing.

6. Store chips at room temperature. Store all reagents and reagent mixes at 4°C when not in use to avoid poor results caused by reagent decomposition. Kit reagents contain dimethyl sulfoxide (DMSO). Because the dye binds to nucleic acids, it should be treated as a potential mutagen and used with appropriate care. Wear hand and eye protection and follow good laboratory practices when preparing and handling reagents and samples. Handle the DMSO stock solutions with particular caution as DMSO is known to facilitate the entry of organic molecules into tissues.

7. Protect dye and dye mixtures from light to prevent their decomposition. Remove light covers only when pipetting, since the dye decomposes when exposed to light and this reduces the signal intensity.

8. The genomic DNA employed for this kind of analyses can be isolated from different plant sources: fresh or dry plant material, roots, shoots, seeds, or other plant organs by using a standard protocol (see ref. 15) or a commercial kit such as DNeasy® Plant Mini Kit (Qiagen) (see ref. 11).

9. 5S-P1 and 5S-P2 NTS primers worked out for many kinds of plants (see refs. 2, 3, 8, 11, 12) without modifying sequence and length, but in some cases it could be necessary to make few modifications in order to increase the specificity.

10. These PCR cycling conditions worked out for a number of plant species (see refs. 3, 8, 11, 12), however, to improve band specificity and intensity, it could be necessary to make few modifications such as increasing number of cycles, annealing temperature, annealing and extension time. The first step to be taken is to run a gradient PCR in order to find the best primer annealing temperature with the template of interest.

11. We always use 2% agarose gel since it shows good resolution for small DNA fragments comprised between 0.2 and 1 kb, the size of the expected PCR products of 5S-rRNA-NTS region.

12. It is good to use a large container, as long as it fits in the microwave, because the agarose boils over easily. The agarose solution can boil over very easily so keep checking it. It is good to stop it after 45 s and give it a swirl. It can become super-heated and not boil until you take it out whereupon it boils out all over your hands. So wear appropriate gloves and hold it at arms length.

13. Allowing the agarose to cool down a little before adding ethidium bromide will minimize the production of ethidium bromide vapor. You can judge about the right temperature when the flask is just not too hot to keep holding in bare hands.

14. After adding the loading buffer you can leave the tip into the microfuge tube. It will be used again to load the sample into the gel.

15. There are lots of different kinds of DNA size markers. Choose a marker with good resolution for the fragment size you expect to see in your sample lanes. For our experiments we use GeneRuler™ 50-bp DNA Ladder (50–1,000 bp) (Fermentas). It is good to load two markers lanes, flanking the samples.

16. UV is carcinogenic and must not be allowed to shine on naked skin or eyes. So wear appropriate eyes and face protection, gloves, and long sleeves.

17. It is necessary to use a sterilized scalpel blade for each sample to avoid cross contamination. Put few layers of plastic wrap under the gel to avoid strong exposure of the DNA to UV light which can damage it. Try to minimize the exposure time while cutting out the required band. Include as little of the surrounding gel as possible.

18. We use half of the reagents for the ligation reaction and we obtain good results anyway.

19. In our experience, incubating reactions overnight increases the efficiency of ligation.

20. Incubating the cells in a shaker at high speed (250 rpm) can decrease the efficiency of transformation. Competent cells can be broken by vigorous shaking. We notice that the incubation of transformation mixes in a circulating water bath at 37°C lead to a good number of colonies on selecting agar plates.

21. If a higher number of colonies are required, spin down the cells ($1,200 \times g$, 2 min) in a microfuge and remove 950 μL of the supernatant. Resuspend the pellet in the remaining 50 μL and plate the solution by following the described procedure.

22. To avoid the formation of star colonies, do not leave the bacteria overgrow.

23. After dissolving the colony in the water, remove the tip and make a small streak in a fresh ampicillin agar plate. From this replica plate, it will be possible to pick up positive colonies for further experiments, such as plasmid preparation.

24. In the colony PCR, it is possible to use the vector universal primers, e.g., SP6, T7, M13 forward and reverse or the 5S-rRNA NTS primers (5S-P1 and 5S-P2) to amplify the inserts harbored by the plasmids. In the first case, larger bands, containing short sequences of the vector will be visible.

25. To pick up bacterial colonies, it is possible to use a sterile loop, a sterile toothpick, or a sterile 200 µL micropipette tip. We advise this last solution since it is faster to use than a loop which must be sterilized on a Bunsen burner each time you pick up a colony and it is better than a toothpick that sometimes can interfere with bacterial growth.

26. We notice that sequencing plasmids with 5S-rRNA-NTS primers, instead than purified bands give better results. This is probably due to the fact that not always 5S-rRNA-NTS primers designed on the conserved regions match perfectly to the DNA sequences. Small differences in matching (not critical for a classical PCR reaction) can lead to poor sequencing results.

27. When possible, design specific primers with a length of 18–20 bp and a good GC content (about 50%) to have high specificity. All desired parameters can be set up on the Primer3 program.

28. Choose restriction enzymes able to discriminate among plant species, varieties, chemotypes, etc. and giving bands of different sizes (see ref. 11).

29. Do not use more enzyme than 10% of the final reaction volume. This is because the enzyme storage buffer contains antifreeze (glycerol) to allow it to survive at –20°C. The glycerol will inhibit the digestion if present in sufficient quantities.

30. Some restriction enzymes require to be incubated at temperatures other than 37°C (e.g., 65°C) (see ref. 11).

31. It is possible to prepare a double digestion reaction, but make sure to use a buffer compatible with both enzymes. Otherwise, you must purify the digestion products of the first enzyme by using a kit such as GFX™ PCR DNA and Gel Band Purification Kit (Amersham Biosciences) to remove reagents before setting up the second reaction (see Fig. 2).

32. Add EDTA and/or heat inactivate the restriction enzyme according to manufacturer's instructions. Restriction endonucleases in combination with nonchelated metal ions may degrade internal DNA markers used in the kit.

33. We notice that allowing the gel-dye mix to equilibrate for more than 30 min (up to 1 h) gives better results.

34. Keep the chip priming station clean and avoid sources of dust or other contaminants while you are running an analysis. Foreign matter in reagents and samples or in the wells of the chip will interfere with assay results.

35. Always insert the pipette tip to the bottom of the well when dispensing the liquid. Placing the pipette at the edge of the well may lead to poor results.

36. We notice that setting the vortex speed at 2,000 rpm gives better results, since less bubbles are created during the vortexing.

37. Use loaded chips within 5 min after preparation, since reagents might evaporate, leading to poor results.

References

1. Cai ZH, Li P, Dong TTX, Tsim KWK (1999) Molecular diversity of 5S-rRNA spacer domain in *Fritillaria* species revealed by PCR analysis. Planta Med 65:360–364

2. Rubiolo P, Matteodo M, Bicchi C et al (2009) Chemical and biomolecular characterization of *Artemisia umbelliformis* Lam., an important ingredient of the alpine liqueur "Genepì". J Agr Food Chem 57:3436–3443

3. Bertea CM, Azzolin CMM, Bossi S et al (2005) Identification of an EcoRI restriction site for a rapid and precise determination of β-asarone-free *Acorus calamus* cytotypes. Phytochemistry 66:507–514

4. Negi MS, Rajagopal J, Chauhan N et al (2002) Length and sequence heterogeneity in 5S rDNA of *Populus deltoides*. Genome 45:1181–1188

5. Baker WJ, Hedderson TA, Dransfield J (2000) Molecular phylogenetics of *Calamus* (Palmae) and related rattan genera based on 5S nrDNA spacer sequence data. Mol Phylogenet Evol 14:218–231

6. Trontin JF, Grandemange C, Favre JM (1999) Two highly divergent 5S rDNA unit size classes occur in composite tandem array in European larch (*Larix decidua* Mill.) and Japanese larch (*Larix kaempferi* (Lamb.) Carr.). Genome 42:837–848

7. Scoles GJ, Gill BS, Xin ZY et al (1988) Frequent duplication and deletion events in the 5S RNA genes and the associated spacer regions of the *Triticeae*. Plant Syst Evol 160:105–122

8. Sugimoto N, Kiuchi F, Mikage M et al (1999) DNA profiling of *Acorus calamus* chemotypes differing in essential oil composition. Biol Pharm Bull 22:481–485

9. Udovicic F, McFadden GI, Ladiges PY (1995) Phylogeny of Eucalyptus and Angophora based on 5S rDNA spacer sequence data. Mol Phylogenet Evol 4:247–256

10. Foster LM, Kozak KR, Loftus MG et al (1993) The polymerase chain reaction and its application to filamentous fungi. Mycol Res 97:769–781

11. Bertea CM, Luciano P, Bossi S et al (2006) PCR and PCR-RFLP of the 5S-rRNA-NTS region and salvinorin A analyses for the rapid and unequivocal determination of *Salvia divinorum*. Phytochemistry 67:371–378

12. Gnavi G, Bertea CM, Usai M et al (2010) Comparative characterization of *Santolina insularis* chemotypes by essential oil composition, 5S-rRNA-NTS sequencing and EcoRV RFLP-PCR. Phytochemistry 71:930–936

13. Ma XQ, Duan JA, Zhu DY et al (2000) Species identification of *Radix Astragali* (Huangqi) by DNA sequence of NTS 5S-rRNA spacer domain. Phytochemistry 54:363–368

14. Gnavi G, Bertea CM, Maffei M (2010) PCR, sequencing and PCR–RFLP of the 5S-rRNA-NTS region as a tool for the DNA fingerprinting of medicinal and aromatic plants. Flavour Frag J 25:132–137

15. Luciano P, Bertea CM, Temporale G et al (2007) DNA internal standard for the quantitative determination of hallucinogenic plants in plant mixtures. Forensic Sci Int Genet 1:262–266

Chapter 9

ISSR: A Reliable and Cost-Effective Technique for Detection of DNA Polymorphism

Maryam Sarwat

Abstract

With the emergence of more and more molecular markers as useful tools in plethora of population genetic and phylogenetic studies, choice of marker system for a particular study has become mind boggling. These marker systems differ in their advantages and disadvantages, so it is imperative to keep in mind all the pros and cons of the technique while selecting one for the problem to be addressed.

Here, we have shed light on the ISSR (intersimple sequence repeat) technique, as a marker of choice if one wants to go for properties such as reliability, simplicity, cost effectiveness, and speed, in addition to assessing genetic diversity between closely related individuals. We have outlined here the whole methodology of this technique with an example of *Tribulus terrestris* as case study.

Key words: DNA markers, ISSR, AFLP, *Tribulus terrestris*, DNA fingerprinting

Abbreviations

AFLP	Amplified fragment length polymorphism
AP-PCR	Arbitrarily primed PCR
BSA	Bulk segregant analysis
CAPS	Cleaved amplified polymorphic sequences
DAF	DNA amplification fingerprinting
IRAP	Inter-retrotransposon amplified polymorphism
ISSR	Intersimple sequence repeats
RAPD	Random amplification of polymorphic DNA
REMAP	Retro-transposon-microsatellite amplified polymorphism
RFLP	Restriction fragment length polymorphism
SAMPL	Selectively amplified microsatellite polymorphic loci
SCAR	Sequence characterized amplified region
SRAP	Sequence-related amplified polymorphism
SSAP	Sequence-specific amplification polymorphism

Nikolaus J. Sucher et al. (eds.), *Plant DNA Fingerprinting and Barcoding: Methods and Protocols*, Methods in Molecular Biology, vol. 862, DOI 10.1007/978-1-61779-609-8_9, © Springer Science+Business Media, LLC 2012

1. Introduction to DNA-Based Markers

Molecular markers are imperative in the assessment of genetic diversity, ranging from nucleotide level (SNPs) to gene and allele frequencies (genotype information), and devising various germplasm conservation programs. Molecular markers are also helpful in studying population structure, solving taxonomic problems, and assigning plants to their correct taxonomic hierarchies which are crucial for phylogenetic studies. With the help of molecular markers, it has become easy to prioritize germplasm conservation programs and to identify duplicates in a collection. These markers are immensely helpful in authentication of germplasms. The advantages of DNA-based markers over other markers are, they are neutral and independent of any environmental cues or developmental stage (temporally and spatially independent). For various features of DNA-based markers readers can refer to (1).

Broadly, DNA-based molecular markers can be categorized as hybridization based or PCR based. The former employ restriction enzyme-digestion of DNA, followed by hybridization with a known labeled probe, which helps in identification of profile. While PCR-based markers involve primers to amplify DNA sequences. A comprehensive review of most of the DNA-based markers is provided by number of authors (see refs. 1–4). Here, in Fig. 1 we are giving a lineology or time-course study of most of them from their emergence till date.

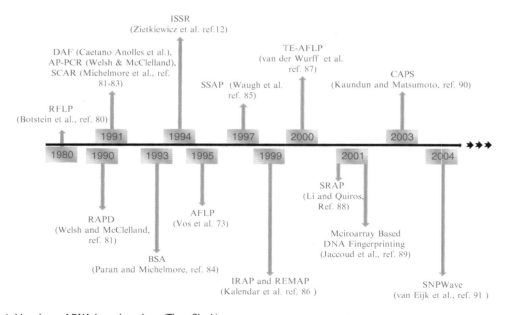

Fig. 1. Lineology of DNA-based markers (Time Clock).

1.1. Microsatellites and ISSR

Microsatellites are tandemly repeated arrays of sequences consisting of di-, tri-, or tetra-nucleotide core units and are ubiquitous, abundant, and highly polymorphic (5, 6) and flanked by highly conserved sequences (7). They are present both in coding and noncoding regions. (AT)n and (GT)n are the most common repeats found in plants with a high frequency of occurrence in the UTRs of coding regions. Closely related species like inbred populations, recently derived or geographically close populations can easily be distinguished by microsatellite-based markers where other molecular tools do not prove useful. Microsatellite primers derived from one species can be used for studying closely related species (8).

High level of allelic variation, presence in genomes of all living organisms, co-dominant way of inheritance, and potential for automated analysis make them excellent molecular markers for genotyping. They are also used for DNA fingerprinting (9), population genetics, and conservation/management of genetic resources (10). The major drawback of microsatellites is the need to isolate them de novo whenever a new species is investigated (10, 11).

Based on the published, unpublished, and in-progress studies that have been conducted using ISSR (intersimple sequence repeat) markers, it is clear that ISSR markers have great potential for studies of natural populations.

Regions of microsatellite abundance "SSR hot spots" are source of ISSR markers. They are also called ISA (inter SSR amplification). First developed by Zietkiewicz et al. (12) for cultivated plants, later find their utility in natural population. ISSR techniques is very much similar to the random amplification of polymorphic DNA (RAPD) technique, except that ISSR primers are designed from microsatellite regions (Fig. 2). They are 16–18-base pair long, 3′ anchored

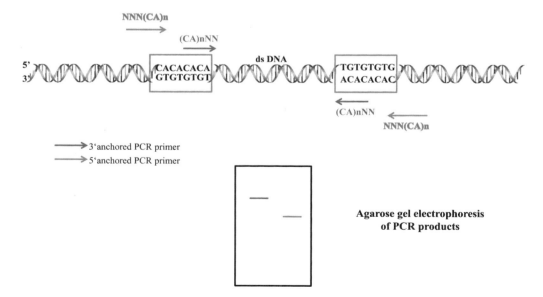

Fig. 2. Schematic representation of ISSR technique.

primers and are able to amplify the inter SSR sequences or sequences flanking SSRs. These regions are 100–3,000-bp long and flanked by closely spaced, inversely oriented microsatellites. The annealing temperatures used in the PCR reaction for ISSR are higher in comparison to RAPDs. Amplified products are anonymous SSR loci. ISSRs are mostly dominant, rarely co-dominant when the length of intervening space between microsatellite has changed. They are economical, fast, and easy to use markers. Their utility has been confirmed in various studies such as detection of somaclonal variation (13), genetic stability (14–16), gene tagging (17, 18), cultivar identification (19), hybrid identification (20) and phylogenetic studies (21, 22), genetic relatedness, (23), and also to determine production quality (24, 25). ISSRs have been employed in genetic diversity studies in *Malus* sp. (26), *Cicer* sp. (27), *Morus alba* (28), *Pisum sativum* (29), *Cynodon* (30), *Lobelia* sp. (31), *Dioscorea alata* (32), wheat (33, 34), *Punica granatum* (35), *Panax ginseng* (36), *Salvia miltiorrhiza* (37), *Terminalia* spp. (38, 39), and *Tribulus terrestris* (40–42). They have also been used for ascertaining the population genetic structure and molecular identification of medicinal plant *Changium smyrnioides* and *Chuanminshen violaceum* (43) and for the analysis of repeat motifs in mungbean (44). They have also been used for molecular characterization of medicinal plant (*Jatropha*, (45)). ISSRs are much more informative than RAPDs (46, 47). Blair et al. (48) detected that ISSR have more polymorphic information content (PIC) than amplified fragment length polymorphism (AFLP).

1.2. Comparison of Molecular Marker Systems

The choice of the marker technology depends upon specific objectives of the study. Each of the different marker systems is associated with advantages and disadvantages with respect to several factors such as:

- Development time and cost
- Capital outlay
- Amount and quality of DNA required
- Robustness
- Informativeness
- Genome coverage
- Reproducibility

Each of these factors can be used independently or in combination to test the efficiency of any particular marker system.

A number of studies have compared the marker systems to judge their relative efficiencies (27, 39, 42, 49–56).

Mignouna et al. (57) compared three PCR-based markers namely, RAPD, AFLP, and SSR using mean number of fragments per assay unit, mean number of fragments per cultivar, number of polymorphic fragments per assay unit, and frequency of polymorphic fragments per assay unit, as measurable criteria. Assay unit is a primer pair for AFLP and a single primer for RAPD. For comparing various techniques, Mignouna et al. (57) utilized the term genotype index (GI) which equals "mean number of genetic profiles per assay unit" divided by "number of genotypes analyzed."

The most commonly used criteria, applicable across all marker systems for comparing efficiencies are PIC, multiplex ratio (MR), and marker index (MI).

Comparison of various marker systems reveals that microsatellite-based markers have the highest information content due to the high allelic diversity present in these sequences. The AFLPs emerged as marker systems having the highest multiplex ratio when compared with RAPD (58, 59), RFLP (60), ISSR (61, 62), and SSR (51, 58).

The utility of SSRs and AFLPs has been combined in SAMPL assays that shows higher marker indices than AFLP (63). Although, both SSR and SAMPL detect polymorphism at microsatellite loci, the latter has a high multiplex ratio than SSR as it generates more number of bands (50 or more) in contrast to only a few (2–5) produced through SSRs.

1.2.1. Polymorphic Information Content

PIC or average heterozygosity (H_{av}) is the probability of detection of polymorphism between two randomly drawn genotypes analyzed in a study or the probability of any solitary band being polymorphic. PIC is calculated for each primer, (64), according to formula:

$$PIC_i = 2f_i(1 - f_i).$$

where, PIC_i is the PIC of the marker i, f_i is the frequency of the amplified allele, and $(1 - f_i)$ is the frequency of the null allele. Average heterozygosity is obtained by taking the average of the PIC value obtained for all the markers and calculated as follows:

$$H_{av} = \sum \left[2f_i(1 - f_i) \right] / N.$$

1.2.2. Multiplex Ratio

Multiplex ratio (MR) can be defined as total number of loci analyzed per assay. Higher the value of MR, better is the performance of marker in polymorphism detection. MR is calculated by dividing total number of bands amplified by total number of primer combinations employed in a particular marker system.

1.2.3. Marker Index

Marker index (MI) shows the overall efficiency of a marker system. MI is the product of PIC and multiplex ratio.

$$MI = H_{av} \times MR.$$

2. Materials

2.1. Experimental Design

Depending upon the problem to be addressed, it is imperative to first design the experiment. For example, if one is going to study population genetics with the help of ISSR markers, first choose the area (areas) to be studied. Then, see which type of population is present, either it is fragmented or well dispersed.

If the problem is to study genetic diversity, then see whether it is interspecific or intraspecific. In case of intraspecific populations, the accessions to be studied should be 6 ft apart in a given area. Otherwise, there are chances of misinterpretation of genetic diversity of the given area.

2.2. Choice/Collections of Samples

Various points to look here are sample should be fresh, as young leaves are a best source of DNA. There should not be any kind of infection in the collected material. After collection, samples should be properly labeled, placed in polybags, and immediately placed in dry ice, for short-term storage. For long-term storage, they should be ideally lyophilized at −40°C and stored at −20°C. The most important part is the proper identification of the sample. Sometimes, plants look so similar to each other leading to misidentification, and resulting in severe consequences. It is better to get them identified by a taxonomist before dwelling into further studies.

For the below-mentioned study, leaf samples of fully grown plants of *T. terrestris* were collected from natural populations growing in Karnataka (six samples from Mysore), Uttaranchal (ten samples from Dehradun), Maharashtra (four samples from Amravati), and Rajasthan (four samples from Udaipur). Fifteen plants growing in the herbal garden at Hamdard University (Jamia Hamdard) represent the genotypes from Delhi. Identification was done on the basis of morphological characters of leaf and fruit and further confirmed by our taxonomist colleague Professor Dr. M.P. Sharma (Department of Botany, Jamia Hamdard). A minimum distance of 30 m was kept between samples.

3. Methods

3.1. Isolation of DNA

Purified DNA of considerable amount is a prerequisite for any kind of fingerprinting studies. The DNA isolation protocol (Doyle and Doyle, ref. 49) given below suits most of the plant types. However, some plants contain more amounts of polysaccharides and polyphenols, for them this protocol does not result in purified DNA. For such kind of plant materials, we have devised our own protocol (see ref. 50).

CTAB DNA isolation protocol of Doyle and Doyle (49): This protocol is also modified by us for optimum yield.

1. Lyophilized leaves (200 mg) are grounded to a fine powder in liquid N_2 using mortar and pestle.

2. Transfer the leaf powder to sterile polypropylene tubes containing 20 ml of prewarmed (65°C) CTAB extraction buffer and mix gently but thoroughly till no clump was visible.

3. Incubate for 40 min at 65°C in a water bath with regular stirring every 5 min.

4. Add equal volume of chloroform:isoamyl alcohol (24:1) and mix thoroughly.

5. Centrifuge at 4,300 × g, at 20°C for 20 min.

6. Measure the upper aqueous phase and transfer to a sterile polypropylene tube.

7. Add RNase A (10 mg/ml) to a final concentration of 100 μg/ml and incubate at 37°C for 30 min.

8. Re-extract the samples with equal volume of chloroform:isoamyl alcohol (24:1) and centrifuge at 2,420 × g, at 20°C for 20 min.

9. Transfer the aqueous phase to a sterile polypropylene tube and measure its volume.

10. Precipitate the DNA by adding equal volume of ice-cold isopropanol to the aqueous phase and incubate at –20°C for 30 min.

11. Precipitate the DNA centrifuged at 6,700 × g, at 4°C for 30 min.

12. Discard the supernatant and wash the DNA pellet with 70% ethanol.

13. Air dry the DNA pellet and dissolve in 200-μl sterile double distilled water.

14. Store at –20°C till further use.

3.2. ISSR Reaction Setup

DNA was diluted to a working concentration of 25 ng/μl. 50 ng DNA from each sample was aliquoted in 0.2-ml PCR tubes. Annealing temperature of primers was according to their T_m. All the components of ISSR analysis were standardized by using various concentrations. Agarose gel concentration for electrophoresis was also standardized. The T_m of the oligos were calculated according to the formula:

$$T_m = 2(A + T) + 4(G + C) - 5°C.$$

To minimize handling and experimental errors, a master-mix comprising of all the reaction components except genomic DNA was assembled for multiple reactions. The assay was carried out in

Table 1
ISSR reaction master mix

Components	Concentration of components	Volume required per reaction (µl)	
DNA (25 ng/µl)	50 ng	2	
10× PCR buffer (with 15 mM MgCl$_2$)	1×	1.5	
25 mM MgCl$_2$	5 mM	1	
10 mM dNTP mix	2 mM each	0.5	
10 µM primer	10 µM	1	
Taq Polymerase (3U/µl)	0.6 U	0.2	
Sterile water	–	8.8	
Total volume	–	15	

15-µl reaction volume and the components were assembled as given in Table 1.

3.3. PCR Amplification   ISSR assay was set up by adding genomic DNA along with the master mix in a PCR tube. The tubes were spun briefly (~10 s). Cycling parameters were following the protocol of Das et al. (51):

One cycle of:

94°C for 2 min.

37°C for 2 min.

72°C for 2 min.

30 cycles of:

94°C for 1 min.

37°C for 1 min (depends upon the T_m of primer used).

72°C for 1 min.

Final extension at 72°C for 10 min.

3.4. Electrophoresis   After the reaction, electrophoresis was performed on 1.4% agarose gel prepared in 0.5× TBE containing 0.5 µg/ml of EtBr at 5 V/cm. It is better to use DNA size marker on each side of the gel, which makes gel scoring less error-prone. Bands were detected under UV light at short wavelength (254 nm) and scored. At least three replicates of each experiment for each primer pair should be performed to nullify misleading results. Only those bands should be considered which are present in all the three gels. Photographs of the gels are captured in UV-transilluminators for further scoring and processing.

3.5. Scoring of Gel

This is the most crucial step of the whole exercise. Bands present in various lanes are compared and scored as 1 and 0 for presence and absence. This binary data is then later fed in MS-excel spreadsheets for further statistical analyses. Here, the rows are equivalent to loci and columns equivalent to taxa.

3.6. Statistical Analysis

The phenetic relationships of the plants are then interpreted for all the ISSR assays using statistical software package NTSYS-pc.

3.6.1. Construction of Binary Matrix by Fragment Matching

The binary matrix data are fed into the NT–Edit program to generate NTSYS-pc compatible file. Comigrating bands were assumed to be originating from the same locus. Only clearly amplified bands were taken for the analysis and the ambiguous ones were omitted.

3.6.2. Calculation of Similarity Matrix Using Jaccard's Coefficient

Jaccard's similarity coefficient was utilized for calculating and representing the pair-wise similarity between operational taxonomic units (OTUs) or genotypes using the formula: $GS_{ij} = a \ / \ a + b + c$.

Where "GS_{ij}" represents genetic similarity between two individuals "i" and "j". "a" is the number of polymorphic bands shared by "i" and "j," "b" is the number of bands present in "i" and absent in "j," and "c" is the number of bands present in "j" and absent in "i" (52).

3.6.3. Cluster Analysis and Construction of Phenetic Dendrogram

Cluster analysis was performed by the distance-based method—SAHN (sequential agglomerative hierarchical nested clustering), where series of successive mergers were used to group individuals. Clustering was done to group individuals with similar characteristics in one group and those with diverse characteristics in the different group. For the graphical representation of these clusters, phenograms were created by employing the Unweighted Pair Group Method of Arithmetical Averages (UPGMA).

Principal correspondence analysis was carried out based on the similarity matrix.

3.7. Mantel Test

Mantel test (53) was used to test the significance of the correlation coefficient between pairs of similarity matrices and for determining cophenetic correlation values using NTSYS-pc (version 2.11 w). Mantel test provides the values for "goodness of fit"; which when above 0.85 suggests goodness of fit (54).

3.8. Bootstrap Analysis

Using WINBOOT software, bootstrap analysis was performed to check the degree of confidence at the nodes of phenograms (55). The original data matrix were used to generate between 100 and 1,000 different variant matrices—through the process of bootstrapping and phenograms, the support values for each node are represented as percentage.

4. Intraspecific Genetic Diversity Analysis of *Tribulus terrestris* Genotypes as Case Study

The molecular marker techniques available for genome scanning are versatile in the context of detection of polymorphism, robustness, reliability, cost effectiveness, amount of labor required as well as the parts of genome they scan. We have compared the molecular marker techniques for their relative efficiencies in detecting polymorphism, average heterozygosity revealed (H_{av}), multiplex ratio (MR), and marker index (MI). Subsequently, we have also checked how concordant are the results obtained from our technique to that from others and the "goodness-of-fit" of each marker system as revealed by their cophenetic correlations by employing Mantel test.

In order to reveal polymorphism in *T. terrestris*, four different assays, namely AFLP, SAMPL, ISSR, and RAPD were employed by us (for detail readers can refer to ref. 42). We used dominant markers here, because of the large number of loci detected per assay, thereby requiring lesser number of assays for a particular study by these markers. Since, single locus markers (RFLP, STMS, STS, and CAPS) detect only a few loci per assay, dominant markers seem to be well suited for diversity analysis.

In this study (42), 21 ISSR primers produced a total of 239 bands with an average of 11.4 amplification products per primer, in the size range of 0.4–2.5 kb. Of 239 bands, 176 were variable, revealing 73.64% polymorphism. The result of ISSR analysis performed for estimating genetic diversity in *T. terrestris* samples is given in Table 2. Maximum number of bands (16) were produced by the primers UBC-4, UBC-5, and S-8; while the least number of bands (5) were generated by the primer UBC-12. Primer UBC-4 revealed 100% polymorphism. The least informative primer was UBC-2 which exhibited only 40% polymorphism (Table 3).

The primer S-13 amplified a total of 13 bands (Fig. 3) in the size range of 0.45–2.3 kb. Of 13 bands, 8 were polymorphic. Some of the monomorphic bands are indicated in figure as "A." "B" denotes a 0.84-kb fragment specific to Mysore genotypes. Another such fragment of 1.25 kb, "C" is specific to Delhi genotypes. "M*" is common across all genotypes except Mysore, and "C*" is common between Dehradun and Amravati regions. "D" is a high intensity band amplified from all genotypes except those from Mysore, which may be the result of multiple repeats, characteristic of microsatellites.

In the following section, the relative efficiency of these marker systems are discussed with respect to detection of polymorphism, pattern of relationship, technicality, economic considerations, and finally reliability and reproducibility. The markers have been compared on the basis of—multiplex ratio (MR), average heterozygosity (H_{av}), marker index (MI), percentage polymorphism detected, and scorability of bands.

Table 2
ISSR analysis for assessment of interzonal genetic diversity in *Tribulus terrestris* (Sarwat et al. (42))

Primer	Total no. of bands	Polymorphic bands	% Polymorphism	Region-specific bands
S-2	7	5	71.43	–
S-3	10	9	90	Delhi—1.8 kb, Mysore—1.3 kb
S-4	6	5	83.33	–
S-5	10	7	70	Dehradun—1.5 kb
S-6	10	7	70	–
S-8	16	12	75	Delhi—0.55 kb
S-11	12	8	66.67	–
S-13	13	8	61.54	Mysore—0.84 kb, Delhi—1.25 kb
S-14	11	7	63.64	Delhi—0.55 kb
S-18	13	10	76.92	Delhi—1.0 kb
UBC-1	8	6	75	Dehradun—0.52 kb
UBC-2	10	4	40	Mysore—0.8 kb
UBC-4	16	16	100	–
UBC-5	16	13	81.25	Delhi—0.72 kb, Mysore—2 kb
UBC-6	13	10	76.92	Delhi—1.35 kb
UBC-7	13	11	84.61	Mysore—0.97 kb
UBC-8	10	5	50	–
UBC-9	12	8	66.67	Delhi—0.82 kb
UBC-10	13	8	61.54	Delhi—0.50 kb
UBC-11	15	14	93.33	Mysore—0.57 kb, Delhi—0.65 kb and 1.30 kb
UBC-12	5	3	60	Mysore—0.872 kb
Total	239	176	73.64	–
(Av.)	–11.4	–8.38		

Among the four marker systems employed, AFLPs have the highest MR and thus proved the best, followed by SAMPL. The multiplex ratio of ISSR is slightly higher than RAPD. Such results are demonstrated in garlic (67) as well. ISSR in *Cicer* sp. have a higher MR (6.3) than RAPD (4.16) (27), thus corroborating my results. Equal MR of AFLP and SAMPL (42) is reported in *Azadirachta indica* (68). High MR of AFLP (43.91), than SAMPL (39.5) and RAPD (6.28) have also been calculated in *Vigna* sp. (69).

Table 3
Comparison of various molecular markers in evaluating genetic diversity of *Tribulus terrestris* (Sarwat et al. (42))

Molecular marker	No. of genotypes	Total no. of bands (n)	Polymorphic bands (p)	Total no. of assays/primer combinations	Multiplex ratio (n/T)	Percentage polymorphic (%p)	Average heterozygosity (H_{av})	Marker index, MI = $H_{av} \times$ MR
AFLP	24	500	437	6	83.33	87.4	0.26	21.66
SAMPL	24	488	471	6	81.33	96.52	0.32	26
ISSR	25	239	176	21	11.38	73.64	0.31	3.52
RAPD	25	276	163	21	13.14	59.06	0.25	3.28

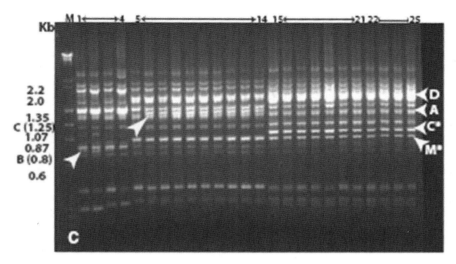

Fig. 3. Representative ISSR fingerprint of *Tribulus terrestris* (Sarwat et al. (42)) utilizing ISSR primer-S-13. *Lanes*—M Marker Standard (mix of λ DNA digested with Hind III and φ×174 DNA digested with Hae III), 1–4 Mysore, 5–14 Delhi, 15–21 Dehradun, and 22–25 Amravati.

Our results (Table 2, detailed results one can have in ref. 42) are in accordance with the findings of Saini et al. (66). They compared the relative efficiency of AFLP, ISSR, and SSR for rice (*Oryza sativa*) and concluded that AFLP markers displayed the highest multiplex ratio. Similarly, Archak et al. (70) obtained high multiplex ratio of AFLP markers as compared to RAPD and ISSR in *Anacardium occidentale*.

In several other studies, MR of AFLP and SAMPL have always been greater than those of RAPD and ISSR, e.g., McGregor et al. (71) recorded MR of 7.8, 10, 10.45, and 124 for SSR, ISSR, RAPD, and AFLP, respectively. Higher MR of AFLP (15.08) than RAPD (0.73) and RFLP (1.47) also recorded in melon (*Cucumis melo*; ref. 72).

The multiplex ratio can, however, be modulated by increasing or decreasing the number of selective nucleotides at the 3′ end of the primers (73). Higher the number of selective nucleotides at the 3 end, higher the stringency and consequently low multiplex ratio. In SAMPL, the MR is additionally dependent on the sequence of microsatellite primers used, as well as the target in the genome.

In comparison to the three marker systems with respect to percent polymorphism, SAMPL provided best results. The capability of SAMPL markers in revealing such high levels of polymorphism is based on the detection of hypervariability in the microsatellite region. Other reports prove that microsatellites do detect high levels of polymorphism (74).

Markers such as SAMPL and ISSR are highly variable, and therefore, become competent to discriminate between closely related genotypes. The hypervariability of SSR is due to variable copy number of repeat units at a locus, which arise due to unequal crossing-over and replication slippage (75).

Saini et al. (65) reported that ISSRs generate much higher percent polymorphism (78.3) than AFLPs (64.0) and SSRs (67.8). Between ISSRs and RAPDs, ISSRs have always performed better than RAPDs. For example, in *Oryza granulata*, Wu et al. (76) demonstrated out-performance of ISSRs over RAPDs (64% vs. 59%) in detecting polymorphism. Similarly, percent polymorphism detected by ISSR (57.7%) was more than what was through RAPD (44.9%) in *Cicer* sp. (27). In *Morus* sp. also Vijayan et al. (77) reported ISSRs to be more polymorphic (80.28%) than RAPDs (75.0%).

The ability to reveal higher polymorphism may be because ISSRs amplify at least two hypervariable microsatellite regions in addition to unique region in between (see Fig. 2).

The third criterion that can be evaluated across markers is of "average heterozygosity." SAMPL revealed highest level of heterozygosity in *T. terrestris* followed by ISSR, AFLP, and RAPD. This can be explained on the basis of hypervariable nature of microsatellites on which both SAMPL and ISSR primers are based.

Tosti and Negri (69) have introduced another criterion termed diversity index (DI) in place of PIC/average heterozygosity (H_{av}). Among the three markers that were compared, highest DI was reported for AFLP (0.24) followed by SAMPL (0.21), and RAPD (0.17). It may be due to higher number of bands amplified by AFLP.

According to my studies (42), RAPD markers did not prove as efficient as the other three marker systems. Among the three, AFLP showed high multiplex ratio while SAMPL and ISSR revealed high average heterozygosity (Table 2). To resolve such situations, Powell et al. (78) identified another criteria termed "marker index" (MI) which determines the overall efficiency of the marker system. In this study (Table 2) with *T. terrestris*, SAMPL markers calculated higher MI (26.0) than AFLP (21.66).

The expression of parameters depends upon the plant system being worked out. This is exemplified by AFLP having higher marker indices than ISSR (61), while Tosti and Negri (69) reported SAMPL as having high MI (1.79) than AFLP (1.62) and RAPD (0.20).

On the basis of the criteria of scorability, AFLP markers come first, as they produced more clear, unambiguous, and distinguishable profiles than any other dominant marker. The SAMPL markers often produced complex profiles with stutter bands that may depend on the nature of region being analyzed (39). Stutter bands were also visible in some ISSR profiles. ISSR and RAPD markers may also generate ambiguous bands of nonhomologous origin. High cophenetic correlation values of AFLP prove the "goodness-of-fit" of these markers, and hence, the robustness of the technique.

The reproducibility of various marker systems has been tested by several groups, e.g., Mc Gregor et al. (71) reported AFLP as

99.6% reproducible, while ISSR are 87% reproducible, and RAPD only 84.4%. Kjolner et al. (65) provided further evidence of high reproducibility of AFLP markers. Vosman et al. (79) have also reported the high correlation ($r = 0.93$) in AFLP results from two different laboratories while examining genetic diversity in rose (*Rosa* sp.).

In conclusion, if one has to go for the properties such as reliability, reproducibility, cost effectiveness, and less labor-intensiveness, then ISSR is the best choice. As each technique is endowed with its advantages and disadvantages, I have outlined the advantages and disadvantages of this technique in the following section.

5. Notes: Advantages and Disadvantages of ISSR Technique

5.1. Advantages

- Major advantage of ISSR is no prior requirement of the genomic sequence.
- Simultaneous assessment of various loci makes the technique fast.
- The PCR products obtained are specific to microsatellite sequences, hence are more reliable than arbitrary sequence-based primer techniques.
- No requirement of cloning and characterization make this technique less time-consuming and economical than other microsatellite-based techniques.
- Effective for distinguishing closely related species.
- Lesser number of steps to be performed make it less cumbersome than AFLP, and hence easy to use.

5.2. Disadvantages

Dominant in nature.

- The ISSR primers sometimes have less specificity to the genome to be scanned, leading to ambiguous fingerprints.
- The poor quality of genomic DNA also leads to poor reproducibility of ISSR results.

Acknowledgment

The research work on *T. terrestris* is funded by a Department of Biotechnology, Government of India grant.

References

1. Sarwat, M., Nabi, G., Das, S., and Srivastava, P. S. (2011) Molecular Markers in Medicinal Plant Biotechnology: Past and Present. *Critical Rev Biotech* (In press) doi:10.3109/07388551.2011.551872.

2. Gupta, P. K., Rustgi, S., and Mir, R. R. (2008) Array-based high-throughput DNA markers for crop improvement. *Heredity* **101**: 5–18.

3. Jones, N., Ougham, H., Thomas, H., and Pašakinskienė, I. (2009) Markers and mapping revisited: finding your gene. *New Phytologist* **183**: 935–966.

4. Arif, I.A., Bakir, M. A., Khan, H. A., Al Farhan, A. H., Al Homaidan, A. A., Bahkali, A. H., Al Sadoon, M., and Shobrak, M. (2010) A Brief Review of Molecular Techniques to Assess Plant Diversity. *Int J Mol Sci* **11**, 2079–2096.

5. Rahman, M. H., Dayanandan, S., and Rajora, O. P. (2000) Microsatellite DNA markers in *Populus tremuloides*. *Genome* **43**, 293–297.

6. Hayden, M. J., and Sharp, P. J. (2001) Targeted development of informative microsatellite (SSR) markers. *Nucleic Acids Res* **29**: e44.

7. Chambers, G. K., and Mac Avoy, E. S. (2000) Microsatellites: consensus and controversy. *Comp Biochem Physiol* **126**, 455–476.

8. Röder, M. S., Plaschke, J., König, U., Börner, A., Sorrells, M., Tanksley, S. D., and Ganal, M. W. (1995) Abundance, variability and chromosomal location of microsatellites in wheat. *Mol Gen Genet* **246**, 327–333.

9. Cordeiro, G. M., Pan, Y-B., and Henry, R. J. (2003) Sugarcane microsatellites for the assessment of genetic diversity in sugarcane gemplasm. *Plant Sci* **165,** 181–189.

10. Zane, L., Bargelloni, L., and Pacarnello, T. (2002) Strategies for microsatellite isolation: a review. *Mol Ecol* **11**: 1–6.

11. Squirrel, J., Hollingsworth, P. M., Woodhead, M., Russell, A., Lowe, A. J., Gibby, M., and Powell, W. (2003) How much effort is required to isolate nuclear microsatellites from plants? *Mol Ecol* **12**, 1339–1348.

12. Zietkiewicz, E., Rafalski, A., and Labuda, D. (1994) Genome fingerprinting by simple sequence repeat (SSR)-anchored polymerase chain reaction amplification. *Genomics* **20**, 176–183.

13. Leroy, X. J., Leon, K., and Branchard, M. (2000) Plant genomic instability detected by microsatellite-primers. *E J Biotech* **3,** No. 2, issue of Aug. 15 http://www.

14. Yuan, X. F., Dai, Z. H., Wang, X. D., and Zhao, B. (2009) Assessment of genetic stability in tissue-cultured products and seedlings of Saussurea involucrata by RAPD and ISSR markers. *Biotechnol Lett* **31**, 1279–87.

15. Zhang, F., Lv, Y., Dong, H., and Guo, S. (2010) Analysis of genetic stability through intersimple sequence repeats molecular markers in micropropagated plantlets of *Anoectochilus formosanus* Hayata, a medicinal plant. *Biol Pharm Bull* **33**, 384–8.

16. Lata, H., Chandra, S., Techen, N., Khan, I. A., ElSohly, M. A. (2010) Assessment of the genetic stability of micropropagated plants of *Cannabis sativa* by ISSR markers. *Planta Med* **76**, 97–100.

17. Ammiraju, J. S. S., Dholakia, B. B., Santra, D. K., Singh, H., Lagu, M. D., Tamhankar, S. A., Dhaliwal, H. S., Rao, V. S., Gupta, V. S., and Ranjekar, P. K. (2001) Identification of inter simple sequence repeat (ISSR) markers associated with seed size in wheat. *Theor Appl Genet* **102**, 726–732.

18. Marczewski, W., Hennig, J., and Gebhardt, C. (2002) The potato virus S resistance gene *Ns* maps to potato chromosome VIII. *Theor Appl Genet* **105**, 564–567.

19. Nagaraju, J., Kathirvel, M., Kumar, R.R., Siddiq, E. A., and Hasnain, S. E. (2002) Genetic analysis of traditional and evolved Basmati and non-Basmati rice varieties by using fluorescence-based ISSR-PCR and SSR markers. *Proc Natl Acad Sci USA* **99**, 5836–5841.

20. Lin, X. C., Lou, Y. F, Liu, J., Peng, J. S., Liao, G. L., Fang, W. (2010) Crossbreeding of *Phyllostachys* species (Poaceae) and identification of their hybrids using ISSR markers. *Genet Mol Res* **9**, 1398–404.

21. Paris, H. S., Yonash, N., Portnoy, V., Mozes-Daube, N., Tzuri, G., and Katzir, N. (2003) Assessment of genetic relationships in *Cucurbita pepo* (cucurbitaceae) using DNA markers. *Theor Appl Genet* **106**, 971–978.

22. Han, Y., and Wang, H. Y. (2010) Genetic diversity and phylogenetic relationships of two closely related northeast China *Vicia* species revealed with RAPD and ISSR markers. *Biochem Genet* **48**, 385–401.

23. Rajwade, A. V., Arora, R. S., Kadoo, N. Y., Harsulkar, A. M., Ghorpade, P. B., and Gupta, V. S. (2010) Relatedness of Indian flax genotypes (*Linum usitatissimum* L.): an inter-simple sequence repeat (ISSR) primer assay. *Mol Biotechnol* **45**, 161–70.

24. Tamhankar, S., Ghate, V., Raut, A., and Rajput, B. (2009) Molecular profiling of "Chirayat"

complex using Inter Simple Sequence Repeat (ISSR) markers. *Planta Med* **75**, 1266–70.

25. Wu, Y., Shi, H. M., Bao, Z., Wang, M. Y., Tu, P. F., and Li, X. B. (2010) Application of molecular markers in predicting production quality of cultivated *Cistanche deserticola*. *Biol Pharm Bull* **33**, 334–9.

26. Goulão, L., and Oliveira, C. M. (2001) Molecular characterisation of cultivars of apple (*Malus×domestica* Borkh.) using microsatellite (SSR, ISSR) markers. *Euphytica* **122**, 81–89.

27. Souframanien, J., and Gopalakrishna, T. (2004) A comparative analysis of genetic diversity in blackgram genotypes using RAPD and ISSR markers. *Theor Appl Genet* **109**, 1687–1693.

28. Awasthi, A. K., Nagaraja, G. M., Naik, G. V., Kanginakudru, S., Thangavelu, K., and Nagaraju, J. (2004) Genetic diversity and relationships in mulberry (genus *Morus*) as revealed by RAPD and ISSR marker assays. *BMC Genetics* **5**: 1 http://biomedcentral.com/1471-2156/5/1

29. Baranger, A., Aubert, G., Arnau, G., Lainé, A. L., Deniot, G., Potier, J., Weinachter, C., Lejeune-Hénaut, I., Lallemand, J., and Burstin, J. (2004) Genetic diversity within Pisum sativum using protein and PCR-based markers. Theor Appl Genet **108**, 1309–1321.

30. Gulsen, O., Sever-Mutlu, S., Mutlu, N., Tuna, M., Karaguzel, O., Shearman, R. C., Riordan, T. P., and Heng-Moss, T. M. (2009) Polyploidy creates higher diversity among *Cynodon* accessions as assessed by molecular markers. *Theor Appl Genet* **118**, 1309–19.

31. Geleta, M., and Bryngelsson, T., (2009) Inter simple sequence repeat (ISSR) based analysis of genetic diversity of *Lobelia rhynchopetalum* (Campanulaceae). *Hereditas* **146**, 122–30.

32. Wu, Z., Leng, C., Tao, Z., Wei, Y., and Jiang, C. (2009) (Genetic diversity of *Dioscorea alata* based on ISSR analysis) *Zhongguo Zhong Yao Za Zhi* 34, 3017–3020.

33. Carvalho, A., Lima-Brito, J., Ma, Ã. Ã. B., and Guedes-Pinto, H. (2009) Genetic diversity and variation among botanical varieties of old Portuguese wheat cultivars revealed by ISSR assays. *Biochem Genet* **47**, 276–94.

34. Thomas, K.G., and Bebeli, P. J. (2010) Genetic diversity of Greek *Aegilops* species using different types of nuclear genome markers. *Mol Phylogenet Evol* **56**, 951–61.

35. Narzary, D., Rana, T. S., and Ranade, S. A. (2010) Genetic diversity in inter-simple sequence repeat profiles across natural populations of Indian pomegranate (*Punica granatum* L.). *Plant Biol* **12**, 806–13.

36. Reunova, G. D., Kats, I. L., Muzarok, T. I., and Zhuravlev, I. N. (2010) (Polymorphism of RAPD, ISSR and AFLP markers of the *Panax ginseng* C. A. Meyer (Araliaceae) genome). *Genetika* **46**:1057–1066.

37. Song, Z., Li, X., Wang, H., and Wang, J. (2010) Genetic diversity and population structure of *Salvia miltiorrhiza* Bge in China revealed by ISSR and SRAP. *Genetica* **138**, 241–9.

38. Sarwat, M. (2010) AFLP, SAMPL, ISSR and RAPD Markers for the Assessment of Genetic Diversity within the Medicinal Herb *Tribulus terrestris*. In: 10th Indo-Pacific Congress on Legal Medicine and Forensic Science (INPALMS), October 25–30, 2010, AMITY University, NOIDA. pp 64–65.

39. Sarwat, M., Das, S., and Srivastava, P.S. (2011) A Comparison of the AFLP and SAMPL Molecular Markers in Characterizing Genetic Diversity of *Terminalia arjuna*- the Backbone of Tasar Silk Industry. *Plant Systemat Evol* **293**, 13–23.

40. Sarwat, M., Das, S., and Srivastava, P. S. (2005a) Genetic diversity analysis of *Terminalia* species through AFLP and SAMPL. In: National Symposium on Plant Biotechnology: New Frontiers, November 18–20, 2005, Lucknow. pp 159.

41. Sarwat, M., Malik, S., Srivastava, T., Narula, A., Das, S., and Srivastava, P. S. (2005b) Studies on genetic diversity of medicinal plants. In: International Conference on Modern Trends in Plant Sciences With Special Reference To The Role of Biodiversity in Conservation. February 17–20, 2005, Amravati, Maharashtra. pp122-123.

42. Sarwat, M., Das, S., and Srivastava, P. S. (2008) Comparison of AFLP, SAMPL, ISSR and RAPD Markers and Analysis of Genetic Diversity in *Tribulus terrestris* Genotypes. *Plant Cell Rep* **27**: 519–528.

43. Qiu, Y. X., Fu, C. X., and Wu, F. J. (2004) Analysis of population genetic structure and molecular identification of *Changium smyrnioides* and *Chuanminshen violaceum* with ISSR marker. *Zhong Yao Cai* **27**, 164–169.

44. Singh, B. M., Sharma, K. D., Katoch, M., Guleria, S., and Sharma, T. R. (2000) Molecular analysis of variability in *Podophyllum hexandrum* Royle an endangered medicinal herb of northwestern Himalaya. *PGR Newsletter* **124**, 57–61.

45. Senthil, Kumar, R., Parthiban, K. T., and Govinda, and Rao, M. (2009) Molecular characterization of *Jatropha* genetic resources through inter-simple sequence repeat (ISSR) markers. *Mol Biol Rep* **36**, 951–6.

46. Korbin, M., Kuras, A., and Zurawicz, E. (2002) Fruit plant germplasm characterisation using molecular markers generated in RAPD and ISSR-PCR. *Cell Mol Biol Lett* **7**, 785–794.

47. Galvan, M. Z., Bornet, B., Balatti, P. A., and Branchard, M. (2003) Inter simple sequence repeat (ISSR) markers as a tool for the assessment of both genetic diversity and gene pool origin in common bean (*Phaseolus vulgaris* L.). *Euphytica* **132**, 297–301.

48. Blair, M. W., Panaud, O., and McCouch, S. R. (1999) Inter-simple sequence repeat (ISSR) amplification for analysis of microsatellite motif frequency and fingerprinting in rice (*Oryza sativa* L.). *Theor Appl Genet* **98**, 780–792.

49. Doyle, J. J., and Doyle, J. L. (1990) Isolation of plant DNA from fresh tissue. *Focus* **12**, 13–15.

50. Sarwat, M., Negi, M., Tyagi, A. K., Lakshmikumaran, M., Das, S., and Srivastava, P. S. (2006) A Standardized Protocol for Genomic DNA Isolation from *Terminalia arjuna* for Genetic Diversity Analysis. *Elect J Biotechnol* **9**: 1. ejbiotechnology.info/content/vol3/issue2/full/2/index.html.

51. Das, S., Rajagopal, J., Bhatia, S., Srivastava, P. S., and Lakshmikumaran, M. (1999) Assessment of genetic variation within *Brassica campestris* cultivars using amplified fragment length polymorphism and random amplification of polymorphic DNA markers. *J Biosci* **24**, 433–440.

52. Jaccard, P. (1908) Nouvelles recherches sur la distribution florale. *Bull Soc Vaudoise Sci Nat* **44**, 223–270.

53. Mantel, N. A. (1967) The detection of disease clustering and a generalized regression approach. *Cancer Res* 27: 209–220.

54. Rohlf, F. J. (2001) NTSYS-pc numerical taxonomy and multivariate analysis system. Version 5.1. Exeter Publishing Ltd., Setauket, N.Y.

55. Yap, I. V., and Nelson, R. (1995) WinBoot: A program for performing bootstrap analysis of binary data to determine the confidence limits of UPGMA-based dendrograms. IRRI Discussion Paper Series No.14. IRRI, Los Baños, Philippines.

56. Goulao, L., Cabrita, L., Oliveira, C. M., and Leitao, J. M. (2001) Comparing RAPD and AFLP analysis in discrimination and estimation of genetic similarities among apple (*Malus domestica* Borkh.) cultivars. *Euphytica* **119**, 259–270.

57. Qian, W., Ge, S., and Hong, D.-Y. (2001) Genetic variation within and among populations of a wild rice (*Oryza granulata*) from China detected by RAPD and ISSR markers. *Theor Appl Genet* **102**, 440–449.

58. Maguire, T. L., Peakall, R., and Saenger, P. (2002) Comparative analysis of genetic diversity in the mangrove species *Avicennia marina* (Forsk.) Vierh. (avicenniaceae) detected by AFLPs and SSRs. *Theor Appl Genet* **104**, 388–398.

59. Kayis, S.A., Hakki, E. E., and Pinarkara, E. (2010) Comparison of effectiveness of ISSR and RAPD markers in genetic characterization of seized marijuana (*Cannabis sativa* L.) in Turkey. *Afr J Ag Res* **5**, 2925–2933.

60. Muthusamy, S., Kanagarajan, S., and Ponnusamy, S. (2008) Efficiency of RAPD and ISSR markers system in accessing genetic variation of rice bean (*Vigna umbellata*) landraces. *Elect J Biotech* **11**, 3. http://www.ejbiotechnology.info/content/vol11/issue3/full/8/index.html

61. Saini, N., Jain, N., Jain, S., and Jain, R. K. (2004) Assessment of genetic diversity within and among Basmati and non-Basmati rice varieties using AFLP, ISSR and SSR markers. *Euphytica* **140**, 133–146

62. Gostimsky, S. A., Kokaeva, Z. G., and Konovalov, F. A. (2005) Studying plant genome variation using molecular markers. *Russian J Genet* **41**, 378–388.

63. Hou, Y.-C., Yan, Z.-H., Wei, Y.-M., and Zheng, Y.-L. (2005) Genetic diversity in barley from West China based on RAPD and ISSR analysis. *Barley Genet. Newslett* **35**: 9–22.

64. Mignouna, H. D., Abang, M. M., and Fagbemi, S. A. (2003) A comparative assessment of molecular marker assays (AFLP, RAPD and SSR) for white yam (*Dioscorea rotundata*) germplasm characterization. *Ann Appl Biol* **142**, 269–276.

65. Kjølner, S., Såstad, S. M., Taberlet, P., and Brochmann, C. (2004) Amplified fragment length polymorphism versus random amplified polymorphic DNA markers: clonal diversity in *Saxifraga cernua*. *Mol Ecol* **13**: 81–86.

66. Saini, N., Jain, N., Jain, S., and Jain, R. K. (2004) Assessment of genetic diversity within and among Basmati and non-Basmati rice varieties using AFLP, ISSR and SSR markers. *Euphytica* **140**, 133–146

67. Ipek, M., and Simon, P. (2002) Evaluation of genetic diversity among garlic clones using molecular markers: comparison of AFLPs, RAPDs and isozymes. In: Plant and Animal Genome X Meeting. Jan 12–16. University of Wisconsin, USA. http://www.hort.wisc.edu/usdavcru/simon/posters/post5.html.

68. Singh, A., Chaudhury, A., Srivastava, P. S., and Lakshmikumaran, M. (2002) Comparison of

AFLP and SAMPL markers for assessment of intra-population genetic variation in *Azadirachta indica* A. Juss. *Plant Sci* **162**: 17–25.

69. Tosti, N., and Negri, V. (2002) Efficiency of three PCR-based markers in assessing genetic variation among cowpea (*Vigna unguiculata* ssp. *unguiculata*) landraces. *Genome* **45**: 268–275.

70. Archak, S., Gaikwad, A. B., Swamy, K. R. M., and Karihaloo, J. L. (2009) Genetic analysis and historical perspective of Cashew (*Anacardium occidentale* L.) introduction into India. *Genome* **52**: 222–230.

71. McGregor, C. E., Lambert, C. A., Greyling, M. M., Louw, J. H., and Warnich, L. (2000) A comparative assessment of DNA fingerprinting techniques (RAPD, ISSR, AFLP and SSR) in tetraploid potato (*Solanum tuberosum* L.) germplasm. *Euphytica* **113**, 135–144.

72. Garcia-Mas, J., Oliver, M., Gómez-Paniagua, H., and de Vicente, M. C. (2001) Comparing AFLP, RAPD and RFLP markers for measuring genetic diversity in melon. *Theor Appl Genet* **101**, 860–864.

73. Vos, P., Hogers, R., Bleeker, M., Reijans, M., van de Lee, T., Hornes, M., Frijters, A., Plot, J., Peleman, J., Kuiper, M., and Zabeau, M. (1995) AFLP: a new technique for DNA fingerprinting. *Nucleic Acids Res* **23**: 4407–4414.

74. Teulat, B., Aldam, C., Trehin, R., Lepbrun, P., Barker, J. H. A., Arnold, G. M., Karp, A., Baudouin, L., and Rognan, F. (2000) An analysis of genetic diversity in coconut (*Cocos nucifera*) populations from across the geographic range using sequence tagged microsatellites (SSRs) and AFLPs. *Theor Appl Genet* **100**, 764–771.

75. Tautz, D., and Renz, M. (1984) Simple sequences are ubiquitous repetitive components of eukaryotic genomes. *Nucl Acids Res* **12**, 4127–4138.

76. Wu, C.-J., Cheng, Z.-Q., Huang, X.-Q., Yin, S.-H., Cao, K.-M., and Sun, C.-J. (2004) Genetic diversity among and within populatios of *Oryza granulata* from Yunnan of China revealed by RAPD and ISSR markers: implications for the conservation of the endangered species. *Plant Sci* **167**, 35–42.

77. Vijayan, K., Kar, P. K., Tikader, A., Srivastava, P. P., Awasthi, A. K., Thangavelu, K., and Saratchandra, B. (2004) Molecular evaluation of genetic variability in wild populations of mulberry (*Morus serrata* Roxb.). *Plant Breed* **123**, 568–572.

78. Powell, W., Morgante, M., Andre, C., Hanafey, M., Vogel, J., Tingey, S. and Rafalski, A. (1996) The comparison of RFLP, RAPD, AFLP and SSR (microsatellite) markers for germplasm analysis. *Mol Breed* **2**, 225–238.

79. Vosman, B., Visser, D., Rouppe, J., Marinus, V., Smulders, J. M., and Eeuwijk, F. (2004) The establishment of 'essential derivation' among rose varieties, using AFLP. *Theor Appl Genet* **109**: 1718–1725.

Chapter 10

Development of Sequence Characterized Amplified Region from Random Amplified Polymorphic DNA Amplicons

Kalpana Joshi and Preeti Chavan

Abstract

Among the PCR-based markers that are most widely used in molecular genetic studies, SCARs (sequence characterized amplified regions) have the advantage of being less sensitive to the conditions of a standard PCR due to its primer size when compared to RAPD (random amplified polymorphic DNA) and hence are more specific and reproducible. Moreover, SCARs require no radioactive isotopes and detect only a single locus. Here, we describe the development of SCAR from RAPD amplicons.

Key words: SCAR, Polymorphisms, DNA marker, Genotyping

1. Introduction

RAPD (random amplified polymorphic DNA) analysis has been widely used for population genetic studies, genetic mapping, genotyping, and marker-assisted selection of desirable traits. However, RAPD analysis is sensitive to reaction conditions, sometimes leading to nonreproducible results. Further, nonstringent annealing temperatures can result in nonspecific amplification which may not be reproducible thereby undermining their direct use as markers. Therefore, RAPD amplicons are converted to SCAR (sequence characterized amplified region) markers. In this technique, the RAPD marker termini are sequenced and longer primers (22–24 nucleotide bases) are designed for specific amplification of a particular locus (1). SCAR primers being longer in length (usually 18–22 nucleotides) have higher annealing temperatures and thus are specific and reproducible. SCARs are useful in genetic mapping studies (codominant SCARs), map-based cloning, comparative mapping, or homology studies among related species, genotyping, and marker-assisted selection of desired traits. SCAR markers have been applied for species identification in various organisms such as

Nikolaus J. Sucher et al. (eds.), *Plant DNA Fingerprinting and Barcoding: Methods and Protocols*, Methods in Molecular Biology, vol. 862,
DOI 10.1007/978-1-61779-609-8_10, © Springer Science+Business Media, LLC 2012

plants (2–4), insects (5, 6), microbes (7), and animals (8). Here, we describe the development of SCAR markers from RAPD amplicons. Following DNA isolation and RAPD analysis, the amplicon of interest is purified, cloned in a suitable vector, and sequenced to develop SCAR markers as described below.

2. Materials

2.1. RAPD

1. Operon primer Kits A, B, C, D, E, F, G, and H (Primer concentration: 5 pmol/µl).
2. *Taq* polymerase, 3 U/µl (Fermentas).
3. 10 mM dNTP mix: Mix 2.5 mM concentration each of dATP/dCTP/dGTP/dTTP.
4. *Taq* Buffer: 100 mM Tris–HCl (pH 9), 15 mM MgCl$_2$, 500 mM KCl, and 0.1% gelatin.
5. Store the above biochemicals at –20°C.

2.2. Purification of RAPD Product

1. Low melting agarose (Sigma).
2. Wizard SV kit for purification of PCR fragment from gel (Promega).

2.3. A-Tailing of RAPD Product

1. Purified RAPD product.
2. 100 mM dATP.
3. *Taq* polymerase Buffer.
4. *Taq* polymerase (3 U/µl) (Fermentas).

2.4. Ligation Reaction

1. A-tailed product.
2. 2× T4 ligase buffer.
3. T4 ligase.
4. 50 ng/µl PGEM®-T Easy vector (Promega).
5. Sterile deionized water.

2.5. Preparation of Competent Cells

1. *Escherichia coli* XL1-Blue.
2. Luria broth.
3. Glycerol.
4. TSB solution: 10 ml TSB contains 86 µl of 5 M NaCl, 1 g polyethylene glycol (mol. wt 8,000), 500 µl DMSO, 100 µl of 1 M MgCl$_2$, 100 µl of 1 M MgSO$_4$, 1 ml glycerol, 5 ml Luria broth, and sterile deionized water to make up volume to 10 ml.
5. Liquid nitrogen.

2.6. Transformation	1. Competent cells (*E. coli* XL1 Blue).
	2. 5× KCM Buffer: 0.5 M KCl, 0.15 M $CaCl_2$, 0.25 M $MgCl_2$.
	3. Ligation mixture.

2.7. Selection of Clones

1. Luria broth powder (HiMedia).
2. X-gal (2% w/v): Dissolve 20 mg X-gal in 1 ml of dimethyl formamide.
3. IPTG (20% w/v): Dissolve 0.2 g IPTG in 1 ml sterile deionized water.
4. 10 mg/ml Ampicillin (US biologicals).

2.8. Screening for Recombinants

NotI from *Nocardia otidiscaviarum* (recognition sequence: 5′-GC/GGCCGC-3′).
Nuclease-free bovine serum albumin (BSA).
10× Assay buffer B: 100 mM Tris–HCl (pH 8), 1 M NaCl, 100 mM $MgCl_2$, 1 mM DTT.
SP6 promoter primer GAT TTA GGT GAC ACT ATA.
T7 promoter primer TAA TAC GAC TCA CTA TAG GG.

2.9. Sequencing

1. 96-Well reaction plate.
2. Big Dye Terminator sequencing buffer (5X).
3. Sequencing Master mix (ABI).
4. Primers.
5. Purified DNA template (insert).

---

3. Methods

3.1. RAPD

To optimize the PCR conditions for RAPD, set up the PCR using 10–100 ng DNA, 50–300 µM dNTP, 1–2.5 mM $MgCl_2$, 0.5–2 U *Taq* polymerase, and 30–45 cycles and 35–40°C of annealing temperature. The RAPD reaction conditions optimized by us are as given in Table 1 (see Note 1).

1. After setting up the reactions, amplification is performed in a thermocycler as described in Table 2. Operon primers are short oligonucleotides of 10 bp length, hence low annealing temperatures are used for amplification.
2. After completion of amplification place the tubes on ice, add 3 µl gel loading buffer to each tube, mix well by centrifugation for 20 s at 7155 ×*g* and load on a 1.5% agarose gel containing ethidium bromide. Electrophorese at 65 V for 1.5 h and visualize the gel on a UV transilluminator (UVP) (see Note 2).

Table 1
RAPD reaction

Reaction components	Volume (effective concentration)	
DNA template	4–5 µl (25 ng)	
10× *Taq* polymerase buffer A	2.5 µl (1×)	
1 mM dNTP	2.5 µl (100 µM)	
3 U/µl *Taq* polymerase	0.3 µl (0.9 U)	
5 pmol/µl RAPD Primer	2 µl (10 pmol)	
Sterile deionized water	q.s.	

Table 2
Thermal cycling parameters for RAPD reaction

Step		Temperature (°C)	Time (min)
1	Initial denaturation	94	5
2	Denaturation	94	1
3	Annealing	36	1
4	Extension	72	2
Go to step 1 and repeat 45 times			
5	Final extension	72	5
Hold at 4°C			

Select a species specific RAPD fragment for conversion to SCAR (Fig. 1).

3.2. Purification of RAPD Product

1. Set up ten PCR reactions for the selected RAPD fragment.

2. Thermal cycling parameters are as described earlier except for a prolonged final extension step of 15 min at 72°C so as to add an adenine nucleotide overhang at the 3' end of the insert.

3. Load the PCR products on a 1.2% low melting agarose gel containing ethidium bromide. After electrophoresis in TBE buffer, visualize the gel under a long-wavelength UV lamp and slice out the fragment of interest in a minimal volume of agarose using a clean sterile scalpel.

4. Purify the RAPD fragment from the agarose gel using the commercial kit as per the instructions of the manufacturer.

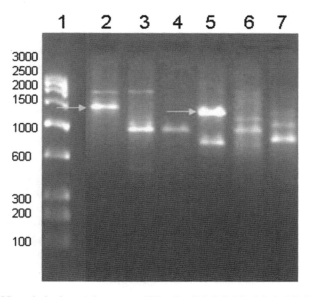

Fig. 1. RAPD analysis. *Lane* 1: Low range DNA ruler (0.1, 0.2, 0.3, 0.6, 1, 1.5, 2, 2.5, and 3 kb). *Lane* 2: *Zingiber officinale* DNA amplified by primer OPC-9. *Lane* 3: *Zingiber zerumbet* DNA amplified by primer OPC-9. *Lane* 4: *Zingiber cassumunar* DNA amplified by primer OPC-9. *Lane* 5: *Z. officinale* DNA amplified by primer OPG-6. *Lane* 6: *Z. zerumbet* DNA amplified by primer OPG-6. *Lane* 7: *Z. cassumunar* DNA amplified by primer OPG-6. *Arrows* indicate putative species-specific RAPD fragments selected for conversion to SCAR.

5. Dissolve the purified DNA in minimum volume of nuclease-free deionized water (18–20 µl) and store at –20°C until further use.

6. Ensure purification of only the desired RAPD fragment by loading on an agarose gel (Fig. 2).

3.3. A-Tailing of RAPD Product

The purified RAPD product is then A-tailed to increase cloning efficiency.

1. The A-tailing reaction is set up in a PCR tube as described in Table 3.

2. Incubate the reaction at 72°C for 20 min in a thermal cycler.

3. After incubation remove the tube from the thermal cycler and store at 4°C.

4. Precipitate the A-tailed RAPD product with two volumes of ethanol and store at –20°C for 30 min.

5. Centrifuged at $21913 \times g$ for 10 min at 4°C to obtain a pellet.

6. Allow the pellet to dry so as to remove ethanol traces and dissolve in 8 µl of sterile deionized water.

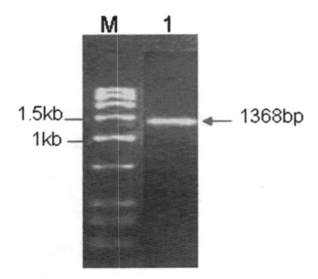

Fig. 2. Purified RAPD fragment. *M*: Low range ruler (0.1, 0.2, 0.3, 0.6, 1, 1.5, 2, 2.5, and 3 kb). *Lane* 1: Purified 1,368-bp RAPD fragment amplified by primer OPC-9.

Table 3
A-tailing reaction

Reaction components		Volume (µl)
Purified RAPD product		87
100 mM dATP		2
Taq polymerase Buffer		10
Taq polymerase (3 U/µl)		1

7. Load 3 µl of the purified product on a 1.5% agarose gel and visualize on a UV transilluminator.

3.4. Ligation Reaction

The purified A-tailed product is ligated with PGEM®-T Easy vector.

1. The ligation reaction is set up as shown in Table 4.
2. Incubate the reaction at 4°C for 12–16 h overnight.

3.5. Preparation of Competent Cells

1. Prepare a 3-ml starter culture of *E. coli* XL1Blue in Luria broth and allow to grow at 37°C for 14–15 h overnight.
2. Add 1% of fresh starter culture to Luria broth, incubate at 37°C shaking at 120 rpm until the O.D.560 = 0.3–0.5 is obtained.

Table 4
Ligation reaction

Reaction components	Volume (μl)
A-tailed product	2–3
T4 ligase	1
2× T4 ligase buffer	5
50 ng/μl PGEM®-T Easy vector	1
Sterile deionized water	q.s.

Table 5
Transformation reaction

Reaction components	Volume (μl)
Competent cells	50
5× KCM	20
Ligation mixture	10
Sterile DI	20

3. Centrifuge at $2795 \times g$ for 5 min at 10°C to pellet down the cells. Discard the supernatant completely.

4. Add 10% glycerol and resuspended the pellet.

5. Centrifuge at 5,000 rpm for 5 min at 10°C. Discard the supernatant and add 5 ml TSB to the pellet. Resuspend the pellet and store on ice for 15 min.

6. Prepare aliquots in precooled microcentrifuge tubes. Snap freeze the tubes immediately by dipping in liquid nitrogen and store at –70°C.

3.6. Transformation

1. Thaw the competent cells on ice.

2. Set up the transformation reaction as shown in Table 5.

3. After mixing the reaction components, store the tube on ice for 30 min and then at room temperature for 10 min.

4. Add 500 μl Luria broth to the tube and shake at 37°C for 2 h.

3.7. Selection of Clones

Luria Agar/ampicillin/IPTG/X-gal plates are prepared as follows:

1. Dissolve 2 g Luria broth powder in 80-ml deionized water and make up the volume to 100 ml.

2. Add 2 g of agar powder and autoclave the medium. Cool the medium to 55°C and add ampicillin to a final concentration of 50 μg/ml. Mix gently and pour 30–35 ml in a sterile plate.

3. After complete solidification of the medium, spread 40 μl of X-gal and 7 μl IPTG stock on the plate. Wrap the plates in aluminum foil and incubate at 37°C for 1–2 h.

4. Innoculate the plates with the transformation reaction mixture and incubate at 37°C for 12–15 h overnight.

5. Identify the recombinant clones by blue/white selection, since the vector is *lacZ* genetically marked.

6. Pick up and inoculate a single white colony in 4-ml Luria broth containing ampicillin (80 μg/ml) and incubate at 37°C shaking at 120 rpm for 12–15 h overnight. Isolate recombinant plasmids by the miniprep method as described by Sambrook and Russel (9).

7. Dissolve the isolated recombinant plasmids in 100 μl of Tris–EDTA buffer.

8. Remove RNA by RNase A treatment as described previously. Finally, dissolve the DNA in 20 μl of Tris–EDTA buffer.

9. Prepare glycerol stocks of the recombinant cells by adding 50 μl of sterile glycerol to 450 μl of culture, mix by vortexing, and freeze immediately by dipping the tubes in liquid nitrogen. Store at −70°C.

3.8. Screening for Recombinants

Screening by restriction endonuclease analysis: Set up restriction digestion reaction (Table 6) with enzymes having sites in the multiple cloning site flanking the insert to check for the presence of desired insert in the purified recombinant plasmids. Incubate the reaction for 4 h at 37°C. Check the size of the restriction digestion product by electrophoresis in 1% agarose gel containing ethidium bromide.

Screening by PCR: Confirm the presence of desired insert by amplification with vector-specific SP6 and T7 promoter primers. The sequences of the primers are given in Table 7. Set up the PCR reactions in a 25 μl volume with reaction conditions same as that for RAPD with the exception of annealing temperature of 55°C and 35 thermal cycles. Analyze the amplification products by electrophoresis in 1.2% agarose gel containing ethidium bromide.

Table 6
Restriction digestion of recombinant plasmids

Reaction components	Volume (μl)
Purified Plasmid DNA	5
10× Buffer B	2
2 U/μl RE	1
1 mg/ml nuclease-free BSA	2

Table 7
Primer sequences

Primer	Sequence 5′–3′
SP6 promoter primer	GAT TTA GGT GAC ACT ATA
T7 promoter primer	TAA TAC GAC TCA CTA TAG GG

Table 8
PCR reaction for sequencing

Reaction components	Volume (μl)
Purified RAPD insert	2
BigDye Terminator Sequencing buffer (5×)	3
Ready Reaction Mix	2
T7/SP6 promoter primer (5 pmol/μl)	1
Deionized water	2

3.9. Sequencing

Sequence the insert from three clones for each RAPD fragment on an ABI 373 automated sequencer (Applied Biosystems, Inc) using ABI BigDye Terminator cycle sequencing kit as per manufacturer's instructions.

1. PCR reaction for sequencing: Set up separate reactions with SP6 and T7 promoter primer as described in Table 8. The thermal cycling conditions for PCR were as described in Table 9.

Table 9
Thermal cycling parameters for sequencing PCR

Step	Temperature	Time
1	96°C	10 s
2	50°C	5 s
3	60°C	4 min
4	Go to step 1 and repeat 25 times	
5	Hold at 4°C	

2. After completion of PCR amplification, prepare the samples for loading into the automated sequencer.

3. Add 80 µl of 80% isopropanol to each reaction and incubate in the dark for 15 min.

4. Centrifuge the plate at 1789 ×*g* for 45 min.

5. After centrifugation add isopropanol and decant by placing the plate in inverted position on a tissue paper.

6. Centrifuge the plate in this position at 14,000 rpm for 1 min.

7. Dry the plate completely and then add 10 µl formamide, incubate at 95°C for 3–5 min followed by incubation on ice for 3–5 min.

8. Load the plates into the automated sequencer. Align the output sequences generated (see Note 3).

9. Analysis of sequence data: Perform homology searches within GenBank's nr database (All GenBank+RefSeq Nucleotides+ EMBL+DDBJ+PDB sequences but no EST, STS, GSS, or phase 0, 1, or 2 HTGS sequences) for all sequenced RAPD fragments using NCBI-BLAST algorithm with the BLASTN 2.2.17 program. Select the unique sequences for use as SCAR markers.

3.10. SCAR Marker Design

Design the SCAR marker sequences by identifying the original 10-bp sequence of the RAPD primer and adding the next approximately 6–10 bp in the DNA sequence. Alternatively, approximately 20-bp sequences can be picked from the sequence using Primer 3 or FastPCR software program. The species-specific SCAR primers are custom synthesized (Integrated DNA Technologies, Inc) (see Note 4).

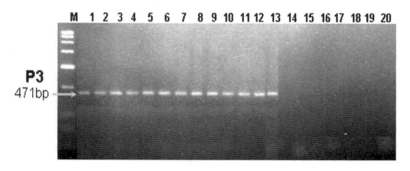

Fig. 3. Species-specific SCAR marker for *Zingiber officinale* Roscoe. *M*: Low range ruler (0.1, 0.2, 0.3, 0.6, 1, 1.5, 2, 2.5, and 3 kb). *Lane 1–13*: *Z. officinale* accessions; *Lane 14–19*: *Zingiber zerumbet* accessions; *Lane 20*: *Z. cassumunar* accession.

3.11. Testing SCAR Primers in Zingiber Species

Test each of the designed SCAR primer pairs (one forward and one reverse) on DNA from the species of interest. A gradient of temperatures in the range of $T_m \pm 5°C$ of the SCAR primers can be used for PCR to determine the optimal annealing temperature. Ensure amplification of the band with exact molecular weight as expected from the sequence data (Fig. 3). Once the optimal annealing temperature for each set of SCAR primer pairs is determined, test them in closely related species to rule out nonspecific amplification.

4. Notes

1. The RAPD reactions are assembled as follows:

   When amplifying several DNA samples with a large number of primers, a master mix containing water, buffer, magnesium chloride, dNTPs, Taq polymerase, and DNA can be prepared and dispensed in PCR tubes. Primers are added to each tube separately. Alternatively, when many DNA samples are to be amplified using one primer, the DNA samples are first added to the PCR tubes followed by a master mix of all other ingredients including the common primer. All reactions are to be set up on ice.

2. Repeat each PCR at least three times and choose only sharp, reproducible bands that are specific for the species of interest for conversion to SCARs.

3. Best sequencing results are obtained approximately 200 bp downstream of the sequencing primers. Internal primers may be designed based on the initial sequence read-outs obtained using SP6 and T7 promoter primers and can be used to sequence the ends of each RAPD fragment. Also for large

fragments that are not sequenced completely using the SP6 and T7 promoter primers, internal nested primers may have to be designed based on the initial sequence read outs.

4. Five to six different primer pairs may be designed for each RAPD fragment that amplify different smaller sized amplicons from the cloned RAPD fragment. Among these the primer pair that amplifies a species-specific and reproducible fragment is chosen. This is especially useful if the RAPD fragment is large and the developed SCAR marker is to be amplified from DNA isolated from a botanical formulation or dried powder where the DNA is likely to be fragmented. In such cases designing primers that amplify smaller amplicons (<500 bp) is useful.

References

1. Paran I, Michelmore RW (1993) Development of reliable PCR-based markers linked to downy mildew resistance genes in lettuce. Theor Appl Genet 85: 985–993

2. Chavan P, Warude D, Joshi K, et al (2008). Development of SCAR marker as a complementary tool for identification of *Zingiber officinale* Roscoe from crude drug and multi-component formulation. Biotech Appl Biochem 50:61–69

3. Lee MY, Doh EJ, Park CH, et al (2006) Development of SCAR marker for discrimination of *Artemisia princeps* and *A. argyi* from other *Artemisia* herbs. Biol Pharm Bull 29:629–633

4. Warude D, Chavan P, Joshi K, et al (2006) Development and application of RAPD-SCAR marker for identification of *Phyllanthus emblica* LINN. Biol Pharm Bull 29: 2313–2316

5. Kethidi DR, Roden DB, Ladd TR, et al (2003) Development of SCAR markers for the DNA-based detection of the Asian long-horned beetle, *Anoplophora glabripennis*(Motschulsky). Arch Insect Biochem Physiol 52:193–204

6. Manguin S, Kengne P, Sonnier L, et al (2002) SCAR markers and multiplex PCR-based identification of isomorphic species in the *Anopheles dirus* complex in Southeast Asia. Med Vet Entomol 16:46–54

7. Trebaol G, Manceau C, Tirilly Y, et al (2001) Assessment of the genetic diversity among strains of *Xanthomonas cynarae* by randomly amplified polymorphic DNA analysis and development of specific characterized amplified regions for the rapid identification of *X. cynarae*. Appl Environ Microbiol 67:3379–3384

8. Yau FC, Wong KL, Wang J, et al. Generation of a sequence characterized amplified region probe for authentication of crocodilian species. J Exp Zool. 2002; 294:382–386

9. Sambrook J, Russell DW(2001) Molecular Cloning: A laboratory manual. Cold Spring Harbor Laboratory Press, New York.

Chapter 11

Authentication of Medicinal Plants by SNP-Based Multiplex PCR

Ok Ran Lee, Min-Kyeoung Kim, and Deok-Chun Yang

Abstract

Highly variable intergenic spacer and intron regions from nuclear and cytoplasmic DNA have been used for species identification. Noncoding internal transcribed spacers (ITSs) located in 18S-5.8S-26S, and 5S ribosomal RNA genes (rDNAs) represent suitable region for medicinal plant authentication. Noncoding regions from two cytoplasmic DNA, chloroplast DNA (*trn*T-F intergenic spacer region), and mitochondrial DNA (fourth intron region of *nad7* gene) are also successfully applied for the proper identification of medicinal plants. Single-nucleotide polymorphism (SNP) sites obtained from the amplification of intergenic spacer and intron regions are properly utilized for the verification of medicinal plants in species level using multiplex PCR. Multiplex PCR as a variant of PCR technique used to amplify more than two loci simultaneously.

Key words: Single-nucleotide polymorphism, Internal transcribed spacers, *trn*T-F region, *nad7* gene, Multiplex PCR

1. Introduction

Conventional medicinal plants contain dozens of active compounds, often of great complexity, tannins, polysaccharides, mucilages, peptides, polyacetylenic alcohols, fatty acids, and so on. They seem to modulate and modify the effects of any "active principles". The clearest example of this is the usage of the term "adaptogenic" to describe the multiple nonspecific effects, such as ginseng. We cannot exclude the possible side effects of medicinal plants, but the demands for them are increasing in the name of "well-being" life. However, most of the medicinal plant products on markets are packaged forms of powders or slices, which no longer bear the original morphological features of the plants (1, 2). In this point of

Nikolaus J. Sucher et al. (eds.), *Plant DNA Fingerprinting and Barcoding: Methods and Protocols*, Methods in Molecular Biology, vol. 862, DOI 10.1007/978-1-61779-609-8_11, © Springer Science+Business Media, LLC 2012

view, we realized that the right authentication of medicinal plant is in demand for correct labeling (barcoding) of them and also for the fair-world marketing. DNA barcoding can potentially used not only for the purposes of plant taxonomists/systematists but also for nonprofessional users such as traditional drug producers/managers and customs officers.

Traditional authentication of medicinal plants, which has relied on morphological and histological differences are limited and quite often unreliable. Chemical analysis via examination of secondary metabolites is another way to be tried. However, this method is not recommendable since it requires large quantity of material for the analysis, and the metabolite profiles are significantly affected by environmental growth conditions as well as storage conditions (3). The identification of medicinal plants based on genotype is not influenced by growth stage and environmental conditions of plants (4). Thus, DNA analysis by molecular biological techniques is highly accepted method for the proper identification of various forms of plants. Several molecular biological methods such as random amplified polymorphic DNA (RAPD) (5–7), arbitrarily primed polymerase chain reaction (AP-PCR) (8), sequence characterized-amplified region (SCAR) (9, 10), restriction fragment length polymorphism (RFLP) (11), loop-mediated isothermal amplification (LAMP) (12), amplification fragment length polymorphism (AFLP) (13), and DNA microarray (14) have been described until now.

In this study, we describe general ways to differentiate medicinally important plants in a more reproducible and robust approach by analyzing single-nucleotide polymorphism (SNP). For the SNP analysis, there are several genes to be tested. (1) The coding sequence of rDNA is distantly related among different species and populations, whereas the noncoding internal transcribed spacers (ITSs) are vary widely. Thus, it renders them as a suitable target for the investigation of phylogenetic relationships within the same species (15). (2) The intron and intergenic spacer sequence of chloroplast DNA (cpDNA) is also used for evolutionary and phylogenetic study (16). The cpDNA *trn*T-F region in land plants comprising the *trn*L (UAA) intron and *trn*L (UAA)-*trn*F(GAA) noncoding intergenic spacer is one of the most widely used chloroplast markers for phylogenetic analysis in plants (17). (3) The noncoding sequences (introns or intergenic spacers) of mitochondrial DNA can be used for phylogenetic studies at low taxonomic levels, including intraspecific diversity studies (18, 19). To speed up the analysis time and accuracy, we adapted multiplex PCR (polymerase chain reaction) based on the identified SNP loci, which allows amplifying more than two loci simultaneously.

2. Materials

All solutions were prepared using ultrapure water (deionized water was purified at 18 MΩ cm at 25°C by Milli-Q plus) and analytical grade reagents were used. Enzymes were obtained from the Korea distributors of Roche, GeneAll, and Frementas, and used with the 10× buffer supplied. Oligonucleotides were synthesized from Genotech (South Korea).

2.1. Buffer Compositions

1. Extraction buffer: 200 mM Tris–HCl (pH 7.5), 250 mM NaCl, 25 mM EDTA, 0.5% SDS.
2. TE: 10 mM Tris–HCl (pH 8.0), 1 mM EDTA.
3. 10× Gel loading buffer: 50% glycerol, 1 mM EDTA, 0.25% bromophenol blue, 0.25% xylene cyanol.
4. 1× TBE buffer: 90 mM Tris–HCl (pH 8.3), 90 mM boric acid, 2.5 mM EDTA (Table 1).

3. Methods

3.1. DNA Extraction Method from Medicinal Plants

Collected medicinal plant samples (mostly half air-dried) were ground in liquid nitrogen for the DNA extraction using plant DNA isolation mini kit (GeneAll, General Biosystem, Korea). In case of fresh leaf tissue with small amount of availability, fresh plant leaf discs were collected in sterile eppendorf tubes containing 300 µl of extraction buffer (29), incubated for 30–60 min at RT. The leaf discs were ground roughly using a plastic macerator and mixed well by vortexing. The same volume of phenol:chloroform (1:1, pH 8.0) was added and incubated for 15 min at RT. The resulting suspension was centrifuged down at $13,000 \times g$ (14,000 rpm in a HERAEUS #3754 rotor) for 10 min. The supernatant was transferred into a new eppendorf tube and precipitated with equal volume of isopropanol. Precipitated genomic DNA was centrifuged down at $13,000 \times g$, washed with ice-cold 70% EtOH, and air dried for 10 min. The pellet was redissolved in 200 µl of TE and 30–50 ng of DNA was taken for the PCR reaction. The concentration of RNA was measured using a GE NanoVue spectrophotometer (GE Healthcare Bio-science, Sweden).

3.2. Preparation of Agarose Gel for the Electrophoresis

Electrophoresis through agarose gel is a standard method used to separate and purify DNA fragments. The gel was solidified with 0.5× TBE buffer containing 5 µl of ethidium bromide (5 mg/ml). DNA fragments were mixed with 1/10 of 10× Gel loading buffer and separated on the agarose gel. Electrophoresis was performed at 5 V/cm using 0.5× TBE buffer.

Table 1
Primer lists used for DNA barcoding of medicinal plants, otherwise other primers are specified in figures

Region	Primer	Nucleotide sequence (5′–3′)	Direction	References
rDNA	ITS5F	5′-GGA AGT AAA AGT CGT AAC AAG G-3′	Forward	(20)
	ITSp1	5′-TAC CGA TTG AAT GRT CCG-3′	Forward	(20, 21)
	ITS4R	5′-TCC TCC GCT TAT TGA TAT GC-3′	Reverse	(20)
	26SF	5′-AGG CGG AGG GGC GGA TAA TG-3′	Forward	(22)
	26SR	5′-ATC TCA GTG GAT CGT GGC AG-3′	Reverse	(22)
	5SP1	5′-GTG CTT GGG CGA GAG TAG TA-3′	Forward	(23)
	5SP2	5′-TTA GTG CTG GTA TGA TCG CA-3′	Reverse	(23)
cpDNA	*trn*L-Fa	5′-CAT TAC AAA TGC GAT GCT CT-3′	Forward	(24)
	*trn*L-Fc	5′-CGA AAT CGG TAG ACG CTA CG-3′	Forward	(24)
	*trn*L-Fe	5′-GGT TCA AGT CCC TCT ATC CC-3′	Forward	(24)
	*trn*L-Fb	5′-TCT ACC GAT TTC GCC ATA TC-3′	Reverse	(24)
	*trn*L-Fd	5′-GGG GAT AGA GGG ACT TGA AC-3′	Reverse	(24)
	*trn*L-Ff	5′-ATT' TGA ACT GGT GAC ACG AG-3′	Reverse	(24)
	MG1	5′-CTA CTG CAG AAC TAG TCG GAT GGA GTA GAT-3′	Forward	(25)
	MG15	5′-ATC TGG GTT GCT AAC TCA ATG-3′	Reverse	(25)
	MATK5	5′-CGA TCC TTT CAT GCA TT-3′	Forward	(25)
	MATK7b	5′-GTA TTA GGG CAT CCC ATT-3′	Reverse	(25)
	PSBA	5′-CGT AAT GCT CAC AAC TTC CC-3′	Forward	(26)
	H	5′-ACT GCC TTG ATC CAC TTG GC-3′	Reverse	(26)
mtDNA	*cox*2/1F	5′-TTT TCT TCC TCA TTC TKA TTT-3′	Forward	(27)
	*cox*2/1 R	5′-CCA CTC TAT TGT CCA CTT CTA-3′	Reverse	(27)
	*cox*2/2 F	5′-GRG TTT ACT ATG GTC AGT GC-3′	Forward	(28)
	*cox*2/2 R	5′-TAG RAA CAG CTT CTA CGA CG-3′	Reverse	(28)
	*nad*74F	5′-TGT CCT CCA TCA CGA TVT CG-3′	Forward	(28)
	*nad*75R	5′-CCA AAT TCT CCT TTA GGT GC-3′	Reverse	(28)

Single code letters: R = A or G, K = G or T, V = A or C or A.

3.3. Multiplex PCR

Before starting the multiplex PCR, several criteria have to be ensured: (a) the primers should not interact with each other, (b) specifically designed primers are need to be designed, (c) each primer batch should be in similar GC content showing homogeneous molecules, and (d) the concentration of primers should be balanced in a way to give equal amounts of each amplicon in multiplex reaction. The thermal cycler conditions were as follows: 5 min at 94°C followed by 36 cycles at 94°C for 30 s, 54°C (Sheng-Ma)/57°C (Jimo)/61°C (Chunpoong) for 30 s, and 72°C for 90 s (see Note 1).

3.4. Amplification and Purification of PCR Products

By combining of different sets of primers (item 3, Subheading 2.1), amplified products were purified using PCR purification kit (QIAGEN, South Korea). The PCR products were visualized by

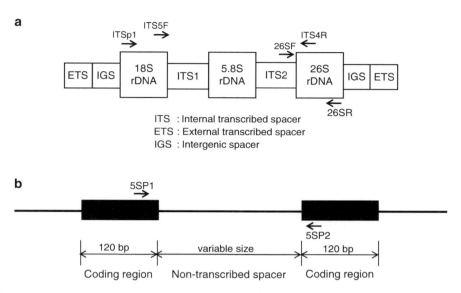

Fig. 1. A schematic diagram of 18S-5.8S-26S rDNA (**a**) and 5S rDNA (**b**). Universal primers are indicated by *arrows*.

the electrophoresis (Mupid-exU) of 10 μl of aliquot on a 1% agarose gel. Separated DNA fragments were visualized on a transilluminator by fluorescence under UV light (254 nm). DNA size marker was loaded to estimate the size of fragments. Purified PCR products were sequenced using the same primers used for the PCR amplification by an automatic DNA sequencer (ABI prism 3700, USA).

3.5. Three Flow Charts of Medicinal Plant Identification by Multiplex-PCR

3.5.1. SNP Analysis by ITS

In eukaryote, ribosomal RNAs (rRNAs) within the nuclear genomes are encoded by two highly conserved multigene families in tandem repeats of three rRNA (18S-5.8S-26S) and 5S rRNA genes (rDNA) (Fig. 1a, b). One unit of 18S-5.8S-26S rDNA contains there coding regions (18S, 5.8S, and 26S rDNA) plus two noncoding ITSs (Fig. 1a). The ITSs are located between the 18S and 5.8S coding regions (ITS1) and between 5.8S and 26S coding regions (ITS2), respectively (30). Although most single-copy genes such as rDNAs in the nuclear genome are not welcomed as barcode (see Notes 2 and 3), the ITS nested 5.8S rDNA was exceptional since the sequences of ITS regions are highly variable (31, 32). The Asian *Cimicifuga* species, known as "Sheng-Ma" in Korea and China, are considered as one of valuable herbs (33). Among four known species (*Cimicifuga heracleifolia*, *Cimicifuga dahurica*, *Cimicifuga foetida*, and *Cimicifuga simplex* DC) which are proven to have medicinal efficacies, species-specific SNP primers were generated based on the sequences of each rDNA containing ITS1 and ITS2 (Fig. 2a). SMh primer was designed to identify *C. heracleifolia*, SMd was to *C. dahurica*, SMs was to *C. simplex*, and SMf was to

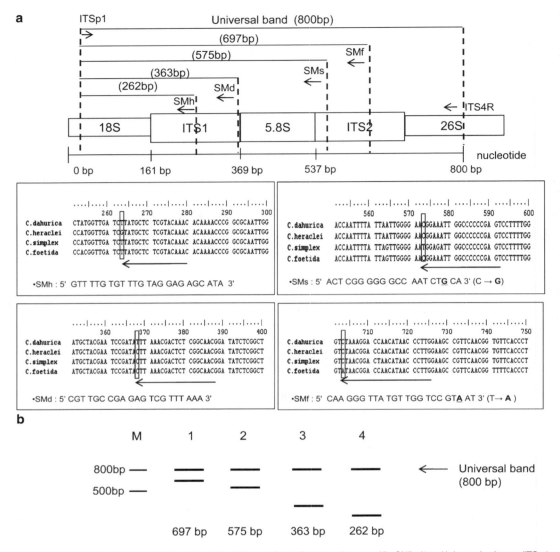

Fig. 2. (**a**) A schematic diagram of 18S-5.8S-26S rDNA and *Cimicifuga* species-specific SNP sites. Universal primers (ITSp1 and ITS4R) and SNP-based primers are indicated by *arrows*. Exact sequences where SNP primers were designed are aligned with SNP sites (*open square box*). Mismatched primer sites are also designated in *parenthesis*. (**b**) Multiplex-PCR shows *Cimicifuga* species-specific bands. *Lane 1*: *Cimicifuga foetida*, *lane 2*: *Cimicifuga simplex*, *lane 3*: *Cimicifuga dahurica*, *lane 4*: *Cimicifuga heracleifolia*. Each 0.25 μM concentration of ITSp1, SMh, SMd, SMf, and SMs primers and 1.25 μM ITS4R were used for the multiplex-PCR.

C. foetida. Primer SMs and SMf are designed by amplification refractory mutation system (ARMS) (34, 35) to enhance allele specificity. Identification of each species was performed by mixing four species-specific primers and two universal primers (ITSp1 and ITS4R) in multiplex PCR reaction. Overall, one universal 800 bp (base pairs) PCR band and additional one more species-specific band for different species were observed (Fig. 2b).

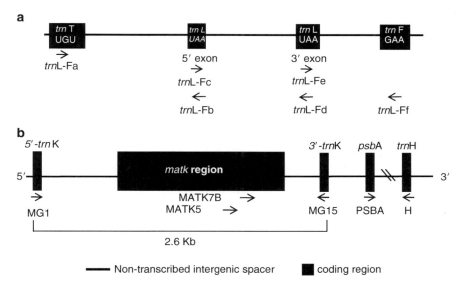

Fig. 3. A schematic diagram of chloroplast DNA *trn*L-F (**a**), *mat*K, and *psb*A-*trn*H regions (**b**). Universal primers per each gene are indicated by *arrows*.

3.5.2. SNP Analysis by Chloroplast DNA

The plastid genome is uniparentally inherited, nonrecombining, and in general, structurally stable. It has been more readily exploited for plant DNA barcoding than nuclear genome, because several sets of the universal primers are available. The most commonly sequenced plastid loci by plant systematists are *rbc*L, *trn*L-F (Fig. 3a), *mat*K, *ndh*F, and *atp*B (36–39). Recently, the spacer region of *psb*A-*trn*H and a portion of the plastid *mat*K gene (Fig. 3b) are reported to be universal DNA barcodes for flowering plants (32, 40). In case of medicinal plant, *trn*T-F intergenic spacer was successfully used for the authentication in species level (Fig. 3a). As an example, molecular identification of medicinal plant *Anemarrhena asphodeloides* Bunge (Jimo) is illustrated (Fig. 4) to discriminate commonly found adulterants of Jimo (2). The available adulterants of Jimo are *Belamcanda chinensis* (Sagan), *Paeonia albiflora* Pall (Chog-chag-yag), and *Peucedanum japonicum* (Shig-ban-pung), which show similar phenotype with Jimo. Intergenic spacer regions of cpDNA from Jimo and Jimo-like samples are sequenced, and observed SNP sites for each sample are used to design species-specific primers; JM9 (*A. asphodeloides* specific), SG9 (*B. chinensis* specific), JJY9 (*P. albiflora* specific), and SBP9 (*P. japonicum* specific) (Fig. 4a). When four species-specific SNP primers and two universal primers for cpDNA (*trn*L-Fc and *trn*L-Ff) were used for multiplex PCR, single 900 bp of universal band with one additional species-specific band were produced (Fig. 4a).

3.5.3. SNP Analysis by Mitochondria DNA

Mitochondrial genome has a remarkable feature of slow-rate evolution (41), resulted in low rate of sequence change. Furthermore, plants rapidly change their mitochondrial genome

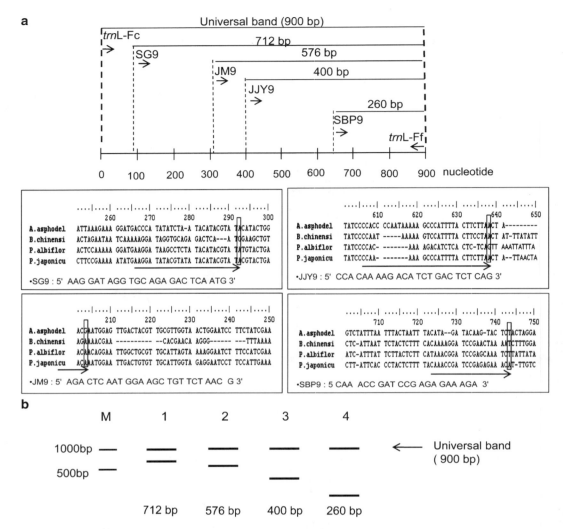

Fig. 4. (a) A schematic diagram of chloroplast DNA trnL-F region and species-specific SNP sites. Universal primers (trnL-Fc and trnL-Ff) and SNP-based primers are indicated by arrows. Exact sequences where SNP primers were designed are aligned with SNP sites (open square box). Anemarrhena asphodeloides Bunge (Jimo) and Jimo-like three other species are indicated. (b) Multiplex-PCR shows Jimo specific bands in lane 2. Lane 1: Belamcanda chinensis, lane 2: Anemarrhena asphodeloides, lane 3: Paeonia albiflora Pall, lane 4: Paeonia japonicum. Each 0.25 μM concentration of trnL-Fc, JM9, SG9, JJY9, SBP9 primers and 0.75 μM trnL-Ff were used for the multiplex-PCR.

structure (42). These features that allow the mitochondrial genes are poor candidates for species-level differentiation (see Note 4). However, the sequence of introns and intergenic regions are highly variable due to the lack of sequence conservation in function (18, 28). The *nad7* gene, encoding the subunit 7 of NADH dehydrogenase complex I, is encoded in the mitochondrial genome in some species of green algae (43) and angiosperms (44), whereas *nad7* gene is encoded in the nuclear genome in fungi (45) and animals (46, 47). Thus mtDNA, *cox2* and *nad7* (Fig. 5a), can also be used

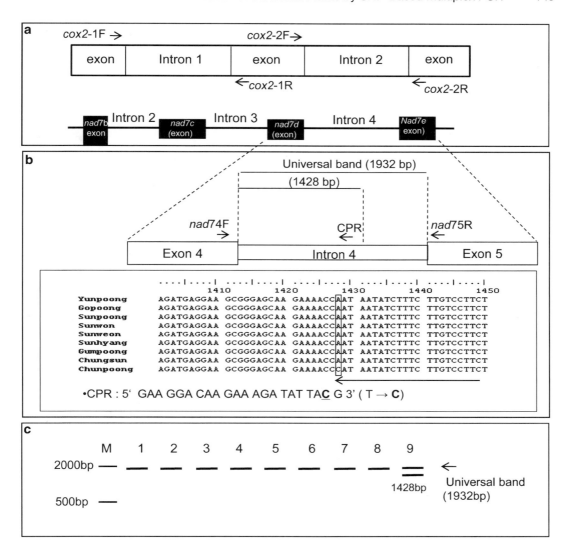

Fig. 5. (**a**) A schematic diagram of mitochondrial DNA *cox2* and *nad7* region and universal primer sites. (**b**) Universal primers (*nad7*4F c and *nad7*5R) and SNP primer (CPR) are indicated by *arrows*. Chunpoong-specific SNP primer (*open square box*) was designed based on the fourth intron sequence of *nad7* gene. Chunpoong-specific mismatched-primer site is designated in *parenthesis*. (**c**) Multiplex-PCR shows Chunpoong-specific bands in *lane 9*. *Lane 1*: Yunpoong, *lane 2*: Gopoong, *lane 3*: Sunpoong, *lane 4*: Sunwon, *lane 5*: Sunweon, *lane 6*: Sunhyang, *lane 7*: Gumpoong, *lane 8*: Chungsun, and *lane 9*: Chunpoong. Each 0.5 µM of universal primers and 0.1 µM of CPR primer were for the multiplex-PCR.

as a useful marker similarly with cpDNA for evolutionary and ecological studies (see Note 5). The fourth intron region of mitochondrial *nad7* gene (Fig. 5a, b) has been successfully used for the identification of Chunpoong among other ginseng cultivars (48). In the fourth intron sequence of *nad7* gene, Chunpoong-specific SNP was identified (Fig. 5b). Most case, only one mismatch at the 3′ termini of designed primer resulted not sufficient for reliable discrimination between alleles. Therefore, additional mismatch next to the 3′-terminal sequence of specific SNP primer was incorporated, T was mismatch to C in CPR primer (Fig. 5b). Multiplex PCR using two universal primers (*nad7*4F and *nad7*5R) and one CPR primer produced universal band (1932 bp) for all ginseng cultivar and an additional Chunpoong-specific band (1,428 bp) only for Chunpoong cultivar (Fig. 5c).

4. Notes

1. PCR optimization is necessary for different types of plant materials by selecting DNA target, designing proper primers, adjusting annealing and elongation temperatures, changing final working concentrations of primers, dNTPs, Mg^{2+}, etc. (2, 33, 48).

2. Since ITS region can be amplified in two smaller fragments (ITS1 and ITS2) adjoining the 5.8S locus, utilization of this region is especially useful for degraded samples. However, the authentication using ITS loci can be resulted in poor success due to the lack of universal primers and the difficulty of gene amplification with low-copy numbers (32). In certain case, only ITS2 region can provide successful outcomes with several reasons: shorter size, easier comparison with other subspecies, and double-checking possibility of sequence errors in aligned secondary structure (49).

3. Smaller amplicon sizes and more copy number promote success with PCR.

4. A short sequence of the mitochondrial CO1 (cytochrome *c* oxidase 1) is a well-known DNA barcode to discriminate other animal species in most groups. However, the divergence rate of CO1 coding region in flowering plants is too low by showing only a few base pairs across 1.4 kb of sequence (50, 51).

5. In some cases, amplification of endophytic or contaminating fungi may occur. This trouble can be eliminated by using well-preserved samples or by using plant-specific primers (52, 53).

Acknowledgment

This work was supported by grants from the Kyung Hee University in 2011 (KHU-20110213) and the Next-Generation BioGreen 21 Program (SSAC, grant #: PJ008204), Rural Development Administration, Republic of Korea (to O.R.L).

References

1. Park, M.J., Kim, M.K., In, J.G., Yang, D.C. (2006) Molecular identification of Korean ginseng by amplification refractory system-PCR. *Food Research International* 39, 568–574.

2. Jigden, B., Wang, H., Narantuya, S. and Yang, D.C. (2010) Molecular identification of oriental plant *Anemarrhena asphodeloides* Bunge ('Jimo') by multiplex PCR. *Mol Biol* 37, 955–960.

3. Yip, T.T., Lau, C.N., But, P.P. and Kong, Y.C. (1985) Quantitative an analysis of Ginsenosides in fresh *Panax Ginseng. The American Journal of Chinese Medicine* 13, 77–88.

4. Zhu, S., Fushimi, H., Cai, S., Komatsu, K. (2004) Species Identification from Ginseng Drugs by Multiplex Amplification Refractory Mutation System (MARMS). *Planta Med* 70, 189–192.

5. Cui, X.M., Lo, C.K., Yip, K.L., Dong, T.T.X., Tsim, K.W.K. (2003) Authentication of *Panax notoginseng* by 5S-rRNA spacer domain and random amplified polymorphic DNA (RAPD) Analysis. *Planta Med* 69, 584–586.

6. Shim, Y.H., Choi, J.H., Park, C.D., Lim, C.J., Cho, J.H., Kim, H.J. (2003) Molecular differentiation of *panax* species by RAPD analysis. *Arch Pharm Res* 26, 601–605.

7. Shaw, P.C., But, P.P. (1995) Authentication of *Panax* species and their adulterants by random-primed polymerase chain reaction. *Planta Med* 61, 466–469.

8. Cheung, K.S., Kwan, H.S., But, P.P., Shaw, P.C. (1994) Pharmacognostical identification of American and Oriental ginseng roots by genomic fingerprinting using arbitrarily primed polymerase chain reaction (AP-PCR). *J Ethnopharmacol* 42, 67–69.

9. Choi, Y.E., Ahn, C.H., Kim, B.B., Yoon, E.S. (2008) Development of species specific AFLP-derived SCAR marker for authentication of *Panax japonicus* C. A. Meyer. *Biol Pharm Bull* 31, 135–138.

10. Wang, J., Ha, W.Y., Ngan, F.N., But, P.P., Shaw, P.C. (2001) Application of sequence characterized amplified region (SCAR) analysis to authenticate Panax species and their adulterants. Planta Med 67, 781–783.

11. Ngan, F., Shaw, P., But, P., Wang, J. (1999) Molecular authentication of *panax* species. *Phytochemistry* 50, 787–791.

12. Sasaki, Y., Komatsu, K., Nagumo, S. (2008) Rapid detection of *Panax ginseng* by loop-mediated isothermal amplification and its application to authentication of ginseng. *Biol Pharm Bull* 31, 1806–1808.

13. Ha, W.Y., Shaw, P.C., Liu, J., Yau, F.C., Wang, J. (2002) Authentication of *Panax ginseng* and *Panax quinquefolius* using amplified fragment length polymorphism (AFLP) and directed amplification of minisatellite region DNA (DAMD). *J Agric Food Chem* 27:50, 1871–1875.

14. Zhu, S., Fushimi, H., Komatsu, K. (2008) Development of a DNA microarray for authentication of ginseng drugs based on 18S rRNA gene sequence. *J Agric Food Chem* 56, 3953–3959.

15. Baldwin, B.G. (1992) Phylogenetic utility of the internal transcribed spacers of nuclear ribosomal DNA in plants: an example from the Compositae. *Mol. Phylogen* 1, 3–16.

16. Olmstead, R.G. and Palmer, J.D. (1994) Chloroplast DNA systematics: a review of methods and data analysis. *Amer. J. Bot* 81, 1205–1224.

17. Quandt, D., Stech, M. (2004) Molecular evolution of the *trn*TUGU *trn*FGAA region in Bryophytes. *Plant Biol* 6(5), 545–554.

18. Hu, D. and Luo, Z. (2006) Polymorphisms of amplified mitochondrial DNA non-coding regions in *Diospyros* spp. *Scientia Horticulturae* 109, 275–281.

19. Deng, K.J., Yang, Z.J, Liu, C., Zhao, W., Feng, J., Ren, Z.L. (2007) Identification and phylogenetic application of unique nucleotide sequence of *nad7* intron 2 in *Rhodiola* (Crassulaceae) species. *Yi Chuan* 29, 371–375.

20. White, T.J., Bruns, T., Lee, S., and Taylor, J. (1990) In, M. A., Innis, D. H. Gelfand, J. J. Sninsky, & T. J. White (Eds.), PCR protocols. A guide to methods and applications: Amplification and direct sequencing of fungal ribosomal RNA genes for phylogenetics. *San Diego: Academic Press* (pp. 315–322)

21. Jigden, B., Wang T., Kim, M.K., Kim, Y.J., In, J.G., Yang, D.K. (2010) Authentication of the Oriental Medicinal Plant *Ligusticum tenuissimum* (Nakai) Kitagawa (Korean Go-Bon) by Multiplex PCR. *Planta Med* 76. 648–651.

22. Kuzoff, R.K., Sweere, J.A., Soltis, D.E., Zimmer, E.A. (1998) The phylogenetic potential of entire 26rDNA sequences in plants. *Mol Biol Evol* 15, 251–263.

23. Wolters, J. and Erdmann, V.A. (1988) Compilation of 5S rRNA and 5S rRNA gene sequences. *Nucleic Acids Research* 16, Suppl. 1–70.

24. Taberlet, P., Gielly, L., Pautou, G. and Bouvet, J. (1991) Universal primer for amplification of three non-coding region of chloroplast DNA. *Plant Molecular Biology* 17, 1105–1109.

25. Khidir, W., Hilu. and Hongping, Liang. (1997) Tne *mat*K gene : Sequence variation and application in plant systematic. *American Journal of Botany* 84(6), 830–839.

26. Saltonstall, K. (2001) A set of primers for amplification of non-coding regions of chloroplast DNA in the grasses. *Molecular Ecology Notes* 1, 76–78.

27. Dumolin-Lapegue, S., Pemonge, M.H., Petit, R.J. (1997) An enlarged set of consensus primers for the study of organelle DNA in plants. *Mol. Ecol* 6, 393–397.

28. Duminil, J., Pemonge, M.H. and Petit, J. (2002) A set of 35 consensus primer pairs amplifying genes and introns of plant mitochondrial DNA. *Mol. Ecol* 2, 428–430.

29. Edwards. K., Johnstone, C. and Thompson, C. (1991) A simple and rapid method for the preparation of plant genomic DNA for PCR analysis. *Nucl Acid Res* 19, 1349.

30. Wendel, J.F., Schnabel, A., Seelanan, T. (1995) Bidirectional inter locus concerted evolution following alloploid speciation in cotton (*Gossypium*). *Proc Natl Acad Sci USA* 92, 280–284.

31. Booy, G., van der Schoot, J., Vosman, B. (2000) Heterogeneity of the internal transcribed spacer 1 (ITS1) in *Tulipa* (Liliaceae). *Plant Sys Evol* 255, 29–41.

32. Kress, W.J., Wurdack, K.J., Zimmer, E.A., Weight, L.A., Janzen, D.H. (2005) Use of DNA barcodes to identify flowering plants. *Proc Natl Acad Sci USA* 102, 8369–8374.

33. Jigden, B., Wang, H., Kim, Y.J., Noh, J.H., Lee, J.I. and Yang, D.C. (2010) Development of a Multiplex Polymerase Chain Reaction Method for Simultaneous Detection of Four *Cimicifuga* Species. *Crop Sci* 50, 1961–1966.

34. Newton, C.R., Graham, A., Heptinstall, L.E., Powell, S.J., Summers, C., Kalsheker, N., Smith, J.C. and Markham, A.F. (1989) Analysis of any point mutation in DNA. The amplification refractory mutation system (ARMS). *Nucleic Acids Res* 17, 2503–2516.

35. In, J.G., Kim, M.K., Lee, O.K., Kim, Y.J., Lee, B.S., Kim, S.Y., Kwon, W.S. and Yang, D.C. (2010) Molecular Identification of Korean Mountain Ginseng Using an Amplification Refractory Mutation System (ARMS). *J. Ginseng Res* 34, 41–46.

36. Gielly, L., and Taberlet, P. (1994) The Use of Chloroplast DNA to Resolve Plant Phylogenies: Noncoding versus *rbc*L Sequences. *Biol. Evol* 11, 769–777.

37. Soltis, D.E., Kuzoff, R.K., Mort, M.E., Zanis, M., Fishbein, M., Hufford, L., Koontz, J. and Arroyo, M.K. (2001) Elucidating deep-level phylogenetic relationships in *Saxifragaceae* using sequences for six chloroplastic and nuclear DNA regions. *Ann. Mo. Bot. Gard* 88, 669–693.

38. Shaw, J., Lickey, E.B., Beck, J.T., Farmer, S.B., Liu, W., Miller, J., Siripun, K.C., Winder, C.T., Schilling, E.E., and Small, R.L. (2005) The Tortoise and the hare II: Relative utility of 21 noncoding chloroplast DNA sequences for Phylogenetic analysis1. *Am. J. Bot* 92, 142–166.

39. Blaxter, M.L. (2004) The Tortoise and the hare II: Relative utility of 21 noncoding chloroplast DNA sequences for Phylogenetic analysis1. *Proc. R. Soc. London* 359, 669–679.

40. Lahaye, R., van der Bank, M., Bogarin, D., Warner, J., Pupulin, F., Gigot, G., Maurin, O., Duthoit, S., Barraclough, T.G. and Savolaine, V. (2008) DNA barcoding the floras of biodiversity hotspots. *Proc Natl Sci USA* 105, 2923–2928.

41. Wolfe, K.H., Li, W.H., and Sharp, P.M. (1987) Rates of nucleotide substitution vary greatly among plant mitochondrial, chloroplast, and nuclear DNAs. *Proc. Natl. Acad. Sci. USA* 84, 9054–9058.

42. Adams, K.L., and Palmer, J.D. (2003) Evolution of mitochondrial gene content: gene loss and transfer to the nucleus. *Mol. Phylogenet. Evol* 29, 380–395.

43. Plante, W.G., I, Lang, B.F., Kuck, U., Burger, G. (1994) Complete sequence of the mitochondrial DNA of the chlorophyte alga *Protheca wickerhamii*. Gene content and genome organization. *J. Mol. Biol* 18, 75–86.

44. Bonen, L., Wiliams, K., Bird, S., Wood, C. (1994) The NADH dehydrogenase subunit 7 gene is interrupted by four group II introns in the wheat mitochondrial genome. *Mol. Gen. Genet* 244, 81–89.

45. Preis, D., van der Pas, J.C., Nehls, U., Rohlen, D.A., Sackmann, U., Jahnke, U., Weiss, H. (1990) The 49 K subunit of NADH: ubiquinone reductase (complex I) frome *Neurospora crassa* mitochondria: primary structure of the gene and the gene and the protein. *Curr. Genet* 18, 59–64.

46. Fearnley, I.M., Runswick, M.J., Walker, J.E. (1993) A homologue of the nuclear coded 49 kb submit of bovine mitochondrial NADH-ubiquinone reductase is coded in chloroplast DNA. *EMBO J* 8, 665–672.

47. Procaccio, V., de Sury, R., Martinez, P., Depetris, D., Rabilloud, T., Soularue, P., Lunardi, J., Issartel, J. (1998) Mapping to

1q23 of the human gene (NDUFS2) encoding the 49-kDa subunit of the mitochondrial respiratory complex I and immunodetection of the mature protein in mitochondria, Mamm. *Genome* 9, 482–484.

48. Wang, H., Hua, S., Kwon, W.S., Jin, H., and Yang, D.C. (2009) Molecular identification of the Korean ginseng cultivar "Chunpoong" using the mitochondrial *nad7* intron 4 region. *Mitochondrial DNA.* 20(2–3); 41–45

49. Chen, S., Yao, H., Han, J., Liu, C., Song, J., Shi, L., Zhu, Y., Ma, X., Gao, T., Pang, X., Luo, K., Li, Y., Li, X., Jia X., Lin, Y., Leon, C. (2010) Validation of the ITS2 Region as a Novel DNA Barcode for Identifying Medicinal Plant Species. *PLoS ONE* 5(1), 8613.

50. Cho, Y., Mower, J.P., Qiu, Y.L. and Palmer, J.D. (2004) Mitochondrial substitution rates are extraordinarily elevated and variable within a genus of flowering plants. *Proc. Natl. Acad. Sci. USA* 101, 17741–17746.

51. Cho, Y., Qiu, Y.L., Kuhlman, P. and Palmer, J.D. (1998) Explosive invasion of plant mitochondria by a group I intron. *Proc. Natl. Acad. Sci. USA* 95, 14244–14249.

52. Zhang, W., Wendel, J.F. and Clark, L.G. (1997) Bamboozled again! Inadvertent isolation of fungal rDNA sequences from bamboos. *Mol. Phylogenet. Evol* 8, 205–217.

53. Cullings, K.W. and Vogler, D.R. (1998) A 5.8S nuclear ribosomal RNA gene sequence database. *Mol. Ecol* 7, 919–923.

Chapter 12

Multiplex PCR Method to Discriminate Artemisia iwayomogi from Other Artemisia Plants

Eui Jeong Doh and Seung-Eun Oh

Abstract

Some plants in the genus *Artemisia* have been used for medicinal purposes. Among them, *Artemisia iwayomogi*, commonly referred to as "Haninjin," is one of the major medicinal materials used in traditional Korean medicine. By contrast, *Artemisia capillaris* and both *Artemisia argyi* and *Artemisia princeps*, referred to as "Injinho" and "Aeyup," respectively, are used to treat diseases different from those for which "Haninjin" is prescribed. Therefore, the development of a reliable method to differentiate each *Artemisia* herb is necessary. We found that a random amplified polymorphic DNA (RAPD) method can be used to efficiently discriminate a few *Artemisia* plants from one another. To improve the reliability of RAPD amplification, we designed primer sets based on the nucleotide sequences of RAPD products to amplify a sequence-characterized amplified region (SCAR) marker of *A. iwayomogi*. In addition, we designed two other primer sets to amplify SCAR markers of "Aeyup" (*A. argyi* and *A. princeps*) along with "Injinho" (*A. capillaris*) and *Artemisia japonica*, which are also traded in Korean herbal markets. Using these three primer sets, we developed a multiplex PCR method concurrently not only to discriminate *A. iwayomogi* from other *Artemisia* plants, but also to identify *Artemisia* plants using a single PCR process.

Key words: *Artemisia* plants, Random amplified polymorphic DNA, Sequence-characterized amplified region marker, Multiplex PCR

1. Introduction

In Korean traditional medicinal markets, specific *Artemisia* herbs are distributed interchangeably with other *Artemisia* herbs (1). For example, *Artemisia iwayomogi* (known as "Haninjin" in Korean medicine), which is an important medicinal herb, is distributed as *Artemisia argyi* and *Artemisia princeps*, which are known as "Aeyup." Furthermore, *A. iwayomogi* is occasionally prescribed as medicinal treatment instead of *Artemisia capillaris*. To solve these problems, the development of reliable methods to discriminate *A. iwaymogi* from other *Artemisia* plants is necessary. For a

Nikolaus J. Sucher et al. (eds.), *Plant DNA Fingerprinting and Barcoding: Methods and Protocols*, Methods in Molecular Biology, vol. 862, DOI 10.1007/978-1-61779-609-8_12, © Springer Science+Business Media, LLC 2012

long time, subjective methods based on morphological features, smell, and taste of herbal plants have been used. However, these methods are not always sufficient for discrimination between plants. Therefore, objective methods such as the ones based on the differences in DNA sequences (2) should be developed.

Previously, we found that some species classified into genus *Artemisia* including *A. iwayomogi*, *A. argyi*, *A. princeps*, *A. capillaries*, *A. japonica*, and *A. keiskeana* can be distinguished from one another by random amplified polymorphic DNA (RAPD) (3). However, the RAPD method has some technical weaknesses. Since RAPD amplification is sensitive to PCR conditions such as the variation in temperature and/or concentrations of chemicals and enzymes, the reproducibility of RAPD results tends to be relatively low. To overcome this problem, sequence-characterized amplified region (SCAR) markers were suggested by Paran and Michelmore (4). Therefore, we designed primer sets to amplify the SCAR markers of *A. iwayomogi* to distinguish one *Artemisia* species from another (6). We also designed two other primer sets to amplify the SCAR markers for the discrimination of other *Artemisia* plants prescribed for medicinal purposes in Korea or that are traded in Korean herbal markets (3, 6). Using these three primer sets, we developed a multiplex PCR (5) method which concurrently discriminates *A. iwayomogi* from other *Artemisia* plants and identifies the *Artemisia* plants in a single PCR process. This method is more efficient in terms of PCR running time and the use of isolated genomic DNA, chemicals, and enzymes compared to conventional PCR to amplify each SCAR marker for *Artemisia* plants in each PCR process. The presented contents and figures regarding the design of the primer sets for amplification of the SCAR markers and development of multiplex PCR method are taken from "Application of the Multiplex PCR Method for Discrimination of *Artemisia iwayomogi* from Other *Artemisia* Herbs" (6).

2. Materials

2.1. Plant Materials

1. Seventeen samples of *Artemisia* plants were obtained in provinces of South Korea and China.

   These include five samples of *A. iwayomogi*, five samples of *A. capillaris*, three samples of *A. princeps*, two samples of *A. argyi*, and one sample each of *A. japonica* and *A. keiskeana*.

2. Fresh leaves from *Artemisia* herbs were washed once with 0.05% Sodium hypochlorite (NaClO), then washed with water.

3. After removing the water, the samples were placed into a labeled 50-mL conical tube, frozen with liquid nitrogen, and stored at $-70°C$ in a deep freezer (see Note 1).

**2.2. Reagents
for Isolation
of Genomic DNA**

Prepare all solutions using ultrapure water (prepared by purifying deionized water to attain a sensitivity of 18 MΩ cm at 25°C) and analytical grade reagents. Prepare and store all reagents and solutions of the purification kit (unless indicated otherwise) at room temperature. RNaseA solution should be stored at 4°C.

1. PureGene DNA purification kit (Gentra, USA).
2. 10% Cetyltrimethyl ammonium bromide (CTAB) buffer: 10% CTAB and 0.7 M NaCl.
   Transfer 10 g of CTAB and 40.9 g of NaCl to a beaker. Mix with 700 mL of water and bring the volume to 1 L with water. Store the solution at room temperature (see Note 2).
3. Phenol:chloroform:isoamlyalcohol solution (25:24:1; Sigma, USA). Store at 4°C (see Note 3).
4. Chloroform:isoamlyalcohol solution (24:1; Sigma).
5. Ice-cold 70% ethanol: Transfer 700 mL ethanol to a 1-L beaker and mix with 300 mL ultrapure water.
   Store at –20°C.
6. Ice-cold isopropanol. Store at –20°C.
7. Liquid nitrogen.

**2.3. Reagents
and Instrument
for PCR Amplification**

All process of PCR amplification mixture preparation are performed at 4°C. Store all reagents for the PCRs at –20°C. Prepare all solutions using ultrapure water (prepared by purifying deionized water to attain a sensitivity of 18 MΩ cm at 25°C).

1. 5 mM Nonspecific UBC primers (University of British Columbia, Canada) for PAPD amplification: Dissolve the freeze-dried RAPD primers in ultrapure water.
2. 10 pmol of the primers designed to amplify SCAR markers: Dissolve the freeze-dried SCAR marker primers in ultrapure water according to the molecular weight of each.
3. T-personal cycler (Biometra, Germany).
4. 1 mM dNTPs.
5. 50 ng of genomic DNA.
6. 1 U Taq polymerase and 1× buffer.
7. 1.5 mM $MgCl_2$.

**2.4. Reagents
for Nucleotide
Sequencing
of PCR Products**

1. 5× TBE buffer: 54 g Tris-base, 27.5 g boric acid, 20 mL of 0.5 M EDTA (pH 8.0) (see Note 4).
   Add about 150 mL water to a 1-L graduated beaker. Weigh out 54 g Tris and 27.5 g boric acid and transfer to the beaker. Dissolve with water to a volume 700 mL and add 20 mL of 0.5 M EDTA solution. Bring the volume to 1 L with water and autoclave. Dilute 5× TBE buffer to 0.5× before use. Store at room temperature.

2. 1.5% Agarose gel: Weigh 1.5% agarose up to the volume. Add 0.5× TBE buffer and dissolve by using a microwave. Stain the gel with EtBr (Sigma).

3. LaboPass™ Gel Extraction Kit (Cosmo Genetech, Republic of Korea).

4. LaboPass™ Plasmid Mini Purification Kit (Cosmo Genetech; see Note 5).

5. pGEM-T-Easy vector (Promega, USA).

6. Competent cells (JM109), store at –70°C.

7. X-gal/IPTG/Ampicillin LB agar plates (see Note 6).

3. Methods

Carry out all procedures at room temperature unless otherwise specified.

3.1. Preparation of Genomic DNA

1. 100 mg of fresh leave sample was frozen with liquid nitrogen and ground to a fine power with chilled mortar and pestle (see Note 7).

2. The genomic DNA of each sample was extracted in accordance with the instruction manual for the PureGene DNA purification kit (Gentra, USA) as follows:

- Add 300 μL of lysis buffer to 100 mg fresh leaf sample powder containing 1.5-mL tube.

- Incubate for 1 h at 65°C. During incubation, invert the tube for every 20 min.

- Add 3 μL RNase A solution and incubated at 37°C for 30 min.

- Add 100 μL of the protein precipitation solution, invert several times, and incubate 5 min on ice.

- Centrifuge at $10,000 \times g$ for 10 min and transfer the supernatant to a new 1.5-mL tube.

- Add an equal volume of ice-cold isopropanol and invert several times.

- Centrifuge $10,000 \times g$ for 5 min and discard the supernatant carefully to avoid losing the pellet.

- Add 700 μL ice-cold 70% ethanol and invert several times.

- Centrifuge $10,000 \times g$ for 5 min and discard the supernatant carefully to avoid losing the pellet.

- Dry the sample to completely remove the ethanol.

- Add 100 μL TE buffer to elute the genomic DNA pellet.

3. Add 400 µL 10% CTAB buffer to 100 µL of the purified genomic DNA (see Notes 2 and 8).

4. Incubate in a 68°C water bath for 30 min.

5. Add 400 µL chloroform:isoamlyalcohol solution (24:1, Sigma) and mix by inverting for 5 min.

6. Centrifuge the samples at $10,000 \times g$ for 5 min at 4°C.

7. Transfer 400 µL of the supernatant to a new 1.5-mL tube and add an equal volume of the phenol:chloroform:isoamlyalcohol solution (25:24:1, Sigma) and mix by inverting for 5 min.

8. Centrifuge the samples at $10,000 \times g$ for 5 min at 4°C and transfer 400 µL of the supernatant into a new 1.5-mL tube.

9. Add an equal volume of ice-cold isopropanol with the sample prepared in step 8 and mix several times by inverting (see Note 9).

10. Centrifuge the samples at $10,000 \times g$ at 4°C for 5 min. A white pellet will be formed.

11. Remove all of the supernatant carefully to avoid losing the pellet and add 700 µL ice-cold 70% ethanol and mix vigorously.

12. Centrifuge the samples at $10,000 \times g$ at 4°C for 5 min.

13. Remove all of the ice-cold 70% ethanol solution carefully to avoid losing the pellet and dry the pellet at room temperature for 15 min to remove the remaining ethanol (see Note 10).

14. Dissolve the dried pellet in 100 µL TE solution.

3.2. Artemisia Plants Are Discriminated Through RAPD Analysis

1. Dilute the genomic DNA to 10 ng/µL in ultrapure water.

2. Prepare the RAPD PCR mixture at 4°C.

    Transfer 50 ng genomic DNA, 600 nM UBC primer (primer for RAPD PCR), 1× PCR buffer, 1.5 mM $MgCl_2$, and 1 U Taq polymerase into a PCR tube and bring the total volume to 20 µL with ultrapure water. Mix all of the ingredients by pipetting up and down (see Note 11).

3. Set PCR program of T-personal cycler.

    The PCR program consisted of 36 cycles as follows: pre-denaturation for 5 min at 94°C, denaturation for 30 s at 94°C. Annealing was conducted for 30 s at 37°C, and extension was conducted for 1 min at 72°C. A final 5 min reaction step was conducted at 72°C. The lid temperature was set at 99°C.

4. PCR products were separated on a 1.5% agarose gel at 100 V for 35 min, and the gel was stained with EtBr (Sigma) for 15 min. The PCR products appeared as bands (Fig. 1) and were analyzed using MyImage (Seoulin Biotechnology, Republic of Korea).

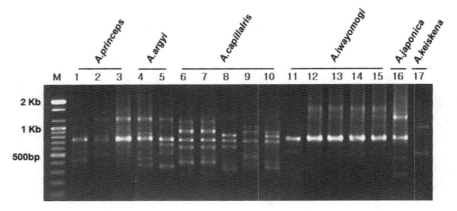

Fig. 1. Polymorphism among *Artemisia* herbs detected by nonspecific UBC primer 391. *M*, 100 bp ladder.

3.3. Development of SCAR Markers

1. Excise each PCR product amplified by nonspecific primers such as UBC primer 391 and that appeared uniquely in specific *Artemisia* plants such as *A. iwayomogi*. Transfer the excised bands to a new 1.5-mL tube (see Note 12).

2. Extract the DNA in each PCR product in accordance with the instruction manual for the LaboPass™ Gel Extraction Kit (Cosmo Genetech, Republic of Korea) as follows:

   • Weigh the gel slice in a 1.5-mL tube. Add three volumes of buffer GB to one volume of gel.

   • Incubate at 50°C until the agarose gel is completely melted (5–10 min).

   • After the slice has dissolved completely, check that the color of the mixture is yellow.

   • Put the spin column into the collection tube and transfer the completely melted mixture into the spin column.

   • Centrifuge for 1 min at 10,000 × *g*. Remove the spin column from the tube and discard the flow-through from the collection tube. Re-insert the spin column in the collection tube.

   • Add 750 μL of buffer NW to the spin column. Let it stand for 5 min to allow some amount of the wash solution to flow-through the column.

   • Centrifuge for 1 min. Transfer the spin column to a new 1.5-mL tube.

   • Apply 35 μL of buffer EB or dH$_2$O to the center of the membrane in the spin column, let stand for 1 min, and centrifuge for 1 min.

3. Subcloning of PCR product is performed according to the instruction manual of pGEM-T-Easy vector (Promega) as follows:

- Prepare the ligation mixture: 50 ng T-Easy vector, 1 U T4 ligase, 1× buffer, and a 1:3 molar ratio of the vector to PCR product. Bring the volume to 20 μL with ultrapure water.

- For ligation, incubate the reaction at room temperature for 1 h and transfer into the tube containing competent cells (JM109) (see Note 13).

- Transform the cells by heat shock at 42°C for 90 s and put the tube immediately on ice for at least 2 min.

- Add 400 μL SOC medium and gently mix by inverting, then incubate in a shaking incubator for 1 h at 37°C.

- Take 200 μL of the transformed competent cells, spread onto an X-gal/IPTG/Amp medium plate gently, and incubate for 16 h at 37°C.

4. Check colony formation on the X-gal/IPTG/Amp medium plate. Pick the white colonies and transfer to 3 mL LB medium containing 100 μg/mL Ampicillin. Incubate at 37°C in a shaking incubator for 16 h (see Note 14).

5. Plasmid DNA (cloned PCR products) are extracted in accordance with the instruction manual for the LaboPass™ Plasmid Mini Purification Kit (Cosmo Gentech) as follows:

- Pellet the bacterial culture by centrifuge for 1 min at 10,000×g in a tabletop centrifuge. Discard the supernatant as much as possible.

- Resuspend the pellet of bacterial cells thoroughly in 250 μL lysis buffer S1 (see Note 5).

- Add 250 μL of buffer S2 and mix by inverting (do not vortex) the tube 4–5 times.

- Add 350 μL of buffer S3 and immediately mix by inverting the tube 4–6 times (do not vortex).

- Centrifuge for 10 min.

- Carefully transfer the supernatant to a spin column and centrifuge for 1 min. Remove the spin column, discard the flow-through, and re-insert the spin column to the collection tube.

- Apply 750 μL of buffer PW and centrifuge for 1 min.

- Remove the spin column, discard the flow-through, and re-inset the spin column into the collection tube.

- Centrifuge for an additional 2 min to remove any residual wash buffer. Transfer the spin column to a new 1.5-mL tube.

- Apply 35 μL of buffer EB or dH$_2$O to the center of the membrane in the spin column, let stand for 1 min, then centrifuge for 1 min.

6. The nucleotide sequences of cloned PCR products are determined by Macrogen sequencing service (Republic of Korea).

7. Compare the determined nucleotide sequence of PCR products of a specific *Artemisia* plant such as *A. iwayomogi* amplified using nonspecific RAPD primers such as UBC 391 (Fig. 2).

8. Based on the established nucleotide sequences, some primer sets are designed to amplify the SCAR marker of a specific *Artemisia* plant such as *A. iwayomogi* (Fig. 2).

9. The PCR mixture was prepared as follows: 50 ng genomic DNA, 1.2 pmol each of the forward/reverse primers, 1× PCR buffer, 1.5 mM MgCl$_2$, and 1 U Taq polymerase. Bring the total volume to 20 μL with ultrapure water. Mix all of ingredient by pipetting up and down (see Note 11).

10. The PCR program consists of 36 cycles as follows: pre-denaturation for 5 min at 94°C and denaturation for 30 s at 94°C. Annealing was conducted for 30 s at 55°C and extension was conducted for 30 s at 72°C. A final 5-min reaction step was conducted at 72°C. The lid temperature was set at 99°C.

11. Among designed primer sets, 2 F1 (5′-ACC TCG GAC CTA AAT ACA-3′)/2F3 (5′-TTA TGA TTC ATG TTC AAT TC-3′) sets were shown to amplify PCR products, which appeared as single band on the gel in *A. iwayomogi* samples (Fig. 3).

12. Using the same method mentioned in Subheadings 3.2 and 3.3, the primer sets 354UF3 (5′-CTA GAG GCC GAC GCG GAC-3′)/354UR3 (5′-ATG CTT TTG GCT ATA TGC AGT C-3′) were designed to amplify common SCAR marker of both *A. capillaris* and *A. japonica*. Fb (5′-CAT CAA CCA TGG CTT ATC CT-3′)/R7 (5′-GCG AAC CTC CCC ATT CCA-3′) were also designed to amplify the common marker of both *A. princepas* and *A. argyi*.

3.4. Development of a Multiplex PCR

1. AYF (5′-ACG GAT ATC TCG GCT C-3′)/AYR (5′-GAA CCA TCG AGT TTT TGA AC-3′) primer sets were designed based on the partial nucleotide sequences of the 5.8S rDNA of *Arabidopsis* as an internal standard which could be used to determine the efficiency of PCR amplification.

2. The PCR mixture was prepared as follows: 50 ng genomic DNA, 0.8 pmol of the 2 F1/2 F3 primers, 4 pmol of the 354UF3/354UR3 primers, 0.08 pmol of the Fb/R7 primers, 0.16 pmol the AYF/AYR primers, 1× PCR buffer, 1.5 mM MgCl$_2$, and 2.5 U Taq polymerase (ABene, USA). Bring the

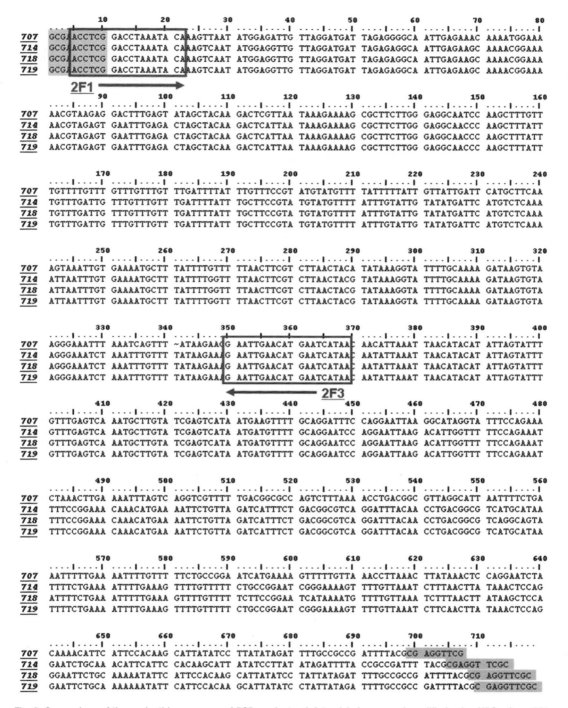

Fig. 2. Comparison of the nucleotide sequence of PCR products of *Artemisia iwayomogi* amplified using UBC primer 391. *Bold boxes* are around nucleotide sequence to indicate a designed primer set. Nucleotides with a *highlighted* background are matching sequences of UBC primer 391.

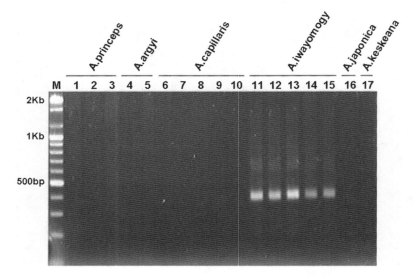

Fig. 3. PCR Products of the designed primer set (2 F1/2 F3) from 17 samples of *Artemisia* herbs. *M*, 100 bp ladder.

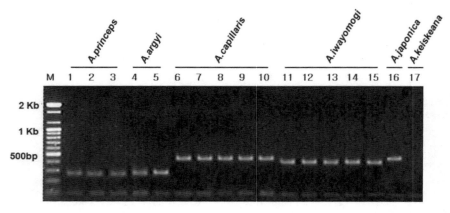

Fig. 4. Multiplex PCR products using four primer sets (2 F1/2 F3, 354UF3.354UR3, Fb/R7, AYF/AYR) from 17 samples of *Artemisia* herbs. AYF/AYR primer set was used as an internal standard of PCR. *M*, 100 bp ladder.

total volume to 25 μL with ultrapure water. Mix all of the ingredients by pipetting up and down (see Note 15).

3. In the multiplex PCR, pre-denaturation was conducted for 12 min at 95°C. Thirty PCR cycles were conducted under the following conditions: denaturation for 30 s at 95°C, annealing for 30 s at 50°C, and extension for 30 s at 72°C. A final 7-min reaction step was conducted at 72°C. The lid temperature was set at 99°C (see Note 15).

4. After the amplified products were separated on a 1.5% agarose gel, the gel was stained with EtBr (Sigma).

5. The amplified products (Fig. 4) were then analyzed using MyImage (Seoulin Biotechnology).

4. Notes

1. Quick-freezing the fresh sample with liquid nitrogen can reduce the damage to the sample tissues and storing in a −70°C freezer increases the storage period of the fresh samples.

2. CTAB and SDS precipitate at temperatures below the room temperature. Therefore, if CTAB/SDS precipitations form, the buffer needs to be warmed prior to use.

3. Before used, check the color of solution. If the color of solution turns to pink or red instead of colorless or yellowish, do not use that. The color change of this solution was cause by phenol oxidization.

4. Add 186.12 g of disodium EDTA.2H$_2$O to 800 mL of ultrapure water. Stir vigorously using a magnetic stirrer. Adjust the pH to 8.0 with NaOH (~20 g of NaOH pellets). Dispense into aliquots and sterilize by autoclaving.

5. After adding the RNase A enzyme to the lysis solution, store at 4°C.

6. Make a stock solution as follows: X-gal dissolved in DMSO at a 20 mg/mL concentration, IPTG dissolved at 100 mM, and ampicillin dissolved at 100 ng/mL in ultrapure water. The dilute stock solution final concentrations are as follows: 20 μg/mL X-gal, 100 μM IPTG, and 100 μg/mL Ampicillin into 15% Bacto-agar containing autoclaved LB medium. Dilute 20 mL in each Petri dish and pre-incubated for 10 h at 37°C.

7. All materials that come into contact with the frozen sample must be chilled with liquid nitrogen before using.

8. CTAB buffer has a viscosity which causes the problem to take a buffer with pipette. Pre-heating the 10% CTAB buffer makes it easier to aliquot CTAB buffer into a new 1.5-mL tube by reducing the viscosity of CTAB buffer.

9. Do not mix violently as this may damage the genomic DNA.

10. Ethanol must be completely evaporated. Any residual ethanol may cause problems in subsequent steps.

11. Taq polymerase is added at the end and mixed by pipetting up and down.

12. Do not expose to the gel to UV light for too long when cutting the amplified samples. Prolonged exposure may possibly cause problems with the ligation reaction.

13. If more colonies are needed, perform the ligation at 16°C for 16 h instead of 1 h. Competent cells should be thawed on ice just before to use. Instead of JM109, DH5α cells can also be used.

14. Blue/white colony selection. Blue colonies represent ones containing only the vector while white colonies are ones containing the vector with the PCR product.

15. Affinity of each primer set is different. Therefore, the concentration of each primer set might be different. At the beginning of the multiplex PCR, add the primer sets one by one until a suitable concentration is found.

References

1. Korea Food & Drug Administration (2002) The Korean Herbal Pharmacopoeia. Korea Food & Drug Administration, Seoul

2. McClelland M, Welsh J (1994) DNA fingerprinting by arbitrarily primed PCR. PCR Methods Appl. 4:59–65

3. Mi Young Lee, Eui Jeong Doh, Chae Haeng Park, Young Hwa Kim, Eung Soo Kim, Byong Seob Ko and Seung-Eun Oh (2006) Development of SCAR Marker for Discrimination of Artemisia *princeps* and *A. argyi* from Other *Artemisia* Herbs. Biol. Pharm.Bull. 29(4):629–633

4. Paran I, Michelmore RW (1993) Development of reliable PCR-based markers linked to downy resistance genes in lettuce. Theor. Appl. Genet. 85:985–993

5. Burgart, L., Robinson, R., Heller, M., Wilke, W., Iakoubova, O., and Chevill, J. (1992) Multiplex polymerase chain reaction. Modern Pathol., 5:320–323

6. Mi Young Lee, Eui Jeong Doh, Eung Soo Kim, Yong Wha, Kim, Byong Seob Ko and Seung-Eun Oh (2008) Application of the Multiplex PCR Method for Discrimination of *Artemisia iwayomogi* from Other *Artemisia* Herbs. Biol. Pharm.Bull. 31(4):685–690

Chapter 13

Loop-Mediated Isothermal Amplification for the Detection of Plant Pathogens

Lisa I. Ward and Scott J. Harper

Abstract

Loop-mediated isothermal amplification (LAMP) is a technique involving the use of four to six primers (two inner primers, two outer primers, and two loop primers) and the strand displacement activity of *Bacillus subtilis*-derived (*Bst*) DNA polymerase. The end result of strand displacement and loop formation and synthesis is the single-temperature amplification of a highly specific fragment from a DNA template at a much greater titre than that obtained with polymerase chain reaction. With LAMP, there are several methods to determine a positive reaction. Presented here are three alternative methods: gel electrophoresis, hydroxynaphthol blue colorimetric dye, and the fluorescent intercalating PicoGreen® reagent.

Key words: Loop-mediated isothermal amplification, Strand displacement, End point detection, Hydroxynaphthol blue, Picogreen, Gel electrophoresis

1. Introduction

Loop-mediated isothermal amplification (LAMP) (1, 2) is a recently developed technique used for the specific amplification of nucleic acid templates at a single temperature. LAMP has been used to detect plant pathogenic fungi, bacteria, nematodes, and viruses (3–9, 16).

The basis of the LAMP method is the strand displacement activity of *Bacillus subtilis*-derived (*Bst*) DNA polymerase with specific activity between 60 and 65°C. Many other polymerases, including the bacteriophage Φ29 and *Escherichia coli* DNA polymerase I, display strand displacement activity although at significantly lower temperature optima which exclude their use in a LAMP reaction. LAMP operates through the priming of heat-denatured positive

Nikolaus J. Sucher et al. (eds.), *Plant DNA Fingerprinting and Barcoding: Methods and Protocols*, Methods in Molecular Biology, vol. 862, DOI 10.1007/978-1-61779-609-8_13, © Springer Science+Business Media, LLC 2012

and negative strands with two template-specific inner primers on which the *Bst* polymerase binds and releases single-strand DNA. This is then targeted by outer primers from which template extension, or more simply, second strand synthesis can occur. LAMP differs from normal polymerase chain reaction (PCR) in that the 3′ and 5′ ends of the inner primers bind to two different, although neighbouring, regions within the template. A 4-mer non-specific oligo "hinge" can be included between the primers so that when bound to ssDNA template, forms a "dumbbell-like" shape that initiates further extension. Two additional loop primers trigger secondary strand displacement reactions within the extended template (2, 10). As the reaction proceeds, the DNA template resembles a cauliflower-like shape with multiple copies joined by the loop-forming inner primers, and being displaced by outer and loop primer pairs. The end result of strand displacement, loop formation and synthesis is the single-temperature amplification of a highly specific fragment from a DNA template at far greater titre than an equivalent PCR reaction (1, 2, 9). The use of six primers provides a greater level of target specificity than can be achieved with the two primers normally used in PCR (1); this is both an advantage and a shortcoming of the LAMP process in that polymorphism, particularly, at the 3′ end of the F1 and 5′ end of the B1 loop primers within the target organism can reduce the likelihood of a successful amplification (6). LAMP is, however, less susceptible to interference from non-target DNA in the reaction, comparable to nested PCR (1). There are several methods to determine whether the LAMP reaction is positive or negative for a target sequence. This protocol describes three different end point detection methods for LAMP: (a) gel electrophoresis, (b) colorimetric hydroxynaphthol blue dye (11), and (c) fluorescent intercalating dye, PicoGreen® reagent (8). LAMP end point detection may also be done using other fluorescent intercalating dyes such as SYBR green, by magnesium pyrophosphate accumulation (4, 12, 13), by precipitation with cationic polymers (14), and by lateral flow devices with labelled primers (15). It should be noted that the LAMP procedure is an end point protocol, similar to standard PCR, and does not lend itself to accurate quantification in the manner of a real-time PCR.

2. Materials

2.1. Reaction Mix Components

1. 10× ThermoPol buffer: 20 mM Tris–HCl, 10 mM $(NH_4)_2SO_4$, 10 mM KCl, 2 mM $MgSO_4$, 0.1% Triton X-100, pH 8.8 (New England Biolabs, Beverly, MA). Store at –20°C.

2. 50 mM $MgSO_4$ (Promega, Madison, WI). Store at –20°C.

3. 5 M Betaine solution (Sigma-Aldrich, St Louis, MO). Store at 4°C.

4. 10 mM dNTPs. Store at –20°C.

5. *Bst* DNA polymerase, Large Fragment (New England Biolabs). Store at –20°C.

2.2. Primers

1. Molecular grade water. Store at 4°C.

2. Micro-centrifuge tubes.

3. LAMP primers: 5-µM working solution of outer primers (F3 and B3), 40-µM working solution of inner primers (FIP and BIP), and 20-µM working solution of loop primers (FL and BL). Store at 4°C until reconstituted, then store at –20°C.

2.3. End Point Detection by Gel Electrophoresis

1. 50× Tris acetic acid/EDTA (TAE) stock buffer: Make up a stock solution of 50× TAE as follows: add 242.0 g of 2 M Tris-base, 57.1 ml of glacial acetic acid (5.7%) (v/v), and 18.6 g of 50 mM disodium EDTA (pH 8.0) to 1 L of deionized water and mix to dissolve.

2. 1× Tris acetic acid/EDTA (TAE) working buffer: Dilute 200 ml of 50× TAE in 10 L of deionized water (see Note 1).

3. Agarose powder.

4. SYBR® Safe DNA gel stain x 10,000 concentrate in DMSO (Invitrogen). Store at room temperature.

5. 100-bp marker ladder (Invitrogen). Store at –20°C.

6. 6× Orange-G loading dye: Using deionized water, make up a 10-ml solution by adding 1.5 g (15% solution w/v) Ficoll 400 (Sigma-Aldrich, St. Louis), 0.15 g of 40 mM EDTA (Sigma-Aldrich), 0.03 g (0.25% solution w/v) Orange-G (Sigma-Aldrich). Store at room temperature in a 500 ml Duran bottle. The dye can be aliquoted into 1 ml volumes to make pipetting easier. (see Note 2).

7. Gel comb.

8. Gel cassette.

9. Gel tank.

10. Power pack.

2.4. End Point Detection Using Hydroxynaphthol Blue Dye

1. Hydroxynaphthol blue (HNB) dye (Sigma-Aldrich): Make up a 10-mM solution of HNB by dissolving 0.62 g of dye in 10 ml of sterile molecular grade water. Make 1 ml aliquots in 1.5-ml micro-centrifuge tubes. Store at –20°C.

2.5. End Point Detection Using PicoGreen® Reagent

1. Quant-iT™ PicoGreen® dsDNA reagent (Component A) in DMSO (Invitrogen): Store between 2 and 6°C and protect from light.

3. Methods

Carry out all procedures at room temperature unless otherwise specified.

3.1. Primer Design

1. Primers for LAMP amplification can be designed by importing a selected sequence into the online PrimerExplorer V4 software (Eiken Chemical Co., Tokyo). Follow the software instructions for importing sequence and generating primers (see Note 3). Four primers are designed for LAMP based on six distinct regions of the target sequence, a forward and backward inner primer (FIP/BIP) and a forward and backward outer primer (F3/B3). After the LAMP primer set has been determined, loop primers can be designed to reduce amplification time and improve specificity. The loop primers are designed based on the primer file of the regular primer set (see Note 4).

3.2. Primer Reconstitution

1. Reconstitute each primer in 100 µl of sterile molecular grade water to create working stocks (see Note 5).

2. For the forward and backward outer primers, prepare a 5-µM working solution by diluting 5 µl of primer stock in 95 µl of molecular grade water (see Note 6).

3. For the forward and backward outer primers, prepare a 40-µM working solution by diluting 40 µl of primer stock in 60 µl of molecular grade water (see Note 6).

4. For the loop primers, prepare a 20-µM working solution by diluting 20 µl of primer stock in 80 µl of molecular grade water (see Note 6).

5. Store reconstituted primers solutions at –20°C.

3.3. LAMP

1. LAMP reactions are generally done in volumes of 20 µl or greater (see Note 7). Prepare the reaction mix using the components shown in Table 1. First calculate the volume of each reagent and primer needed for the number of reactions being performed (including a positive and negative control), plus one additional reaction to account for pipetting error. Add the required amount of each reagent into a 1.5-ml micro-centrifuge tube and mix thoroughly (see Note 8).

2. Pipette 23 µl of reaction mix into the required number of 200-µl PCR tubes.

3. Add 2 µl of nucleic acid template into each PCR tube and cap the tubes (see Note 9).

4. Place tubes in a thermocycler (see Note 10) and incubate at 63°C for 60 min (see Note 11), followed by a 2-min enzyme inactivation step at 80°C.

Table 1
Mastermix components for LAMP reaction

Reagents	Volume per reaction (µl)
10× ThermoPol reaction buffer	2.50
50 mM MgSO$_4$ (6–8 mM final)	3.00
5 M Betaine (0.8 M final)	4.00
10 mM dNTPs (1.4 mM final)	3.50
5 µM F3 Primer (0.2 µM final)	1.00
5 µM B3 Primer (0.2 µM final)	1.00
40 µM FIP Primer (1.6 µM final)	1.00
40 µM BIP Primer (1.6 µM final)	1.00
20 µM F-loop Primer (0.8 µM final)	1.00
20 µM B-loop Primer (0.8 µM final)	1.00
Bst DNA Polymerase 8 U/µl	1.00
Sterile molecular grade water	3.00
DNA template	2.00
Total volume	25.00

3.4. End Point Detection by Gel Electrophoresis

1. Mix 1.5 g of agarose gel with 100 ml of 1× TAE in a 500-ml Duran bottle. Cap the bottle loosely and heat in a microwave for approximately 2.5 min until the agar powder has completely dissolved (see Note 12). Leave the heated agar to cool for 5–10 min, then add 2 µl of SYBR® Safe to the agar (see Notes 13 and 14). Swirl the agar in the bottle to distribute the stain. Pour the agar into a gel cassette of appropriate size and immediately insert a gel comb appropriate for the number of samples being tested. Leave the gel to set for approximately 40 min (see Note 15).

2. Once set, place the gel in a gel tank containing enough 1× TAE working buffer to cover the gel by a depth of around 0.5 cm.

3. Load each well with 15 µl of PCR product containing 2 µl 6× Orange-G loading dye.

4. Load 5 µl of marker ladder into the wells on either side of the samples.

5. Run the gel, e.g. 100 V for around 40 min.

6. Remove the gel cassette with gel from the gel tank and drain off any remaining 1× TAE buffer. Remove the gel from the cassette and place on either a UV or blue light trans-illuminator to view LAMP products (Fig. 1) (see Notes 16 and 17).

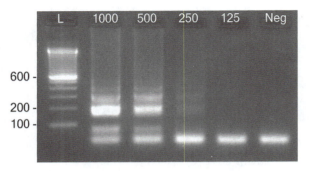

Fig. 1. Results of loop-mediated isothermal amplification (LAMP) of serially diluted *Xylella fastidiosa* DNA (copy numbers of 1,000 to 125) after 60-min incubation at 65°C. Note the ladder pattern of typical of LAMP amplification products. L: Invitrogen 100 bp ladder.

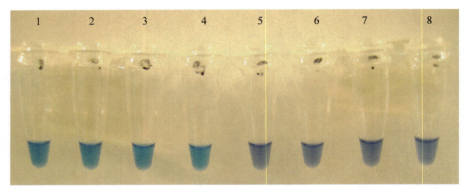

Fig. 2. An example of positive (*blue*, tubes 1–4) and negative (*purple*, tubes 5–8) LAMP reactions for *Xylella fastidiosa* DNA using hydroxynaphthol blue dye as an indicator.

3.5. End Point Detection Using Hydroxynaphthol Blue Dye

1. Follow step 1 in Subheading 3.3 to prepare reaction mix, however, add 0.3 µl of a 10-mM solution of HNB per 25 µl reaction to give a final concentration of 150 µM and reduce the volume of water in the reaction mix accordingly.

2. Follow the steps 2–4 in Subheading 3.3 to set up the LAMP reaction.

3. On completion of the LAMP reaction, look for a colour change from purple to blue, indicative of positive amplification of target nucleic acid (Fig. 2) (see Note 17).

3.6. End Point Detection by PicoGreen® Reagent

1. Follow step 1 in Subheading 3.3 to prepare reaction mix.

2. Follow the steps 2–4 in Subheading 3.3 to set-up the LAMP reaction.

3. On completion of the LAMP reaction, add 2 µl of undiluted PicoGreen® to the 25 µl reaction mix. Look for a colour change from orange to yellow-green, indicative of positive amplification of target nucleic acid (Fig. 3) (see Note 17).

Fig. 3. An example of positive (*yellow-green*) and negative (*orange*) LAMP reactions for *Xylella fastidiosa* DNA using PicoGreen® reagent as an indicator.

4. Notes

1. Tris borate/EDTA (TBE) can be used as an alternative to TAE. 50× TAE or 50× TBE concentrate buffer can be purchased from Bio-Rad Laboratories (Hercules, CA), which can then be used to make a 1× working solution. Stock concentrates may also be available from other suppliers of molecular biology reagents.

2. Loading dyes, bromophenol blue and xylene cyanol are alternatives to Orange-G, and are available from a number of suppliers of molecular biology reagents, e.g. Promega.

3. It is recommended to initially read the online manual which gives details of the key factors in LAMP primer design as well as providing instructions on changing parameters for primer design if required. Primers can be successfully designed using default parameters, however, there are options to change parameters for AT or GC rich sequence.

4. The optimal size for LAMP products is generally under 300 bp. The default settings in the PrimerExplorer V4 software will give a product of around 200 bp.

5. The outer and loop primers can be ordered as desalted, however, it is recommended to use HPLC purified FIP and BIP primers. Depending on the inner primers and the target, these primers may work as desalted, however, they may not be as efficient as HPLC purified.

6. A new LAMP assay should initially be optimised by testing components of the reaction mix. It is recommended to test the following: $MgSO_4$ from 4 to 8 mM; Betaine from 0.6 to 1 M; dNTPs from 0.2 to 1.0 mM of each dNTP (dATP, dTTP, dCTP, dGTP); and *Bst* from 1 to 12 units. Primer concentration should also be optimised with inner primers between 1 and 2 μM, outer primers between 0.2 and 0.5 μM, and loop primers between 0.5 and 1 μM. The concentration shown in

this protocol is such that it allows 1 μl of each primer to be added to the reaction mix.

7. A volume of 20 μl or greater is recommended particularly if using a colour change dye as this gives sufficient volume for the colour change to be visible. In addition, a 20-μl or greater volume is needed to allow all the reagents to be added at the appropriate concentration.

8. LAMP is very sensitive to contamination. To reduce the risk of contamination, the PCR reaction mix should preferably be made inside a clean PCR hood which has not come into contact with nucleic acid. Likewise, the addition of nucleic acid to the reaction mix should be done in a separate area from where the reaction mix reagents are handled.

9. The method used to extract nucleic acid template for the LAMP process varies depending on the source material and whether RNA or DNA is required for the procedure. Commercial silica-based columns, e.g. Qiagen Plant Mini Kit, have been frequently used to extract DNA from different templates. These columns generally provide clean total DNA or RNA preparations with a minimum of inhibitor presence. Examples of where commercial kits have been used to extract nucleic acid for LAMP include microbial cell cultures (8, 17, 18), animal tissue cultures (6), and plants (3, 7). However, a crude CTAB method has also been used to successfully extract the citrus greening organism from *Citrus* species (19). Similar crude heat lysis methods have been applied to bacterial species (5, 20). Immuno-capture RT-LAMP has been successfully applied for virus detection in chrysanthemum samples, suggesting that very simple preparations may be sufficient.

10. If required, the reaction may be incubated in a heat-block or by immersing in a water-bath rather than a thermocycler. However, if a thermocycler with a heated lid is not used, 50 μl of mineral oil may need to be placed over the top of the reaction to stop evaporation occurring.

11. Incubation time and temperature may vary depending on target nucleic acid and primers. To obtain optimal reaction conditions for the detection of target nucleic acid, newly designed LAMP primers should initially be tested over a range of temperatures, e.g. 60–65°C and time, e.g. 15–120 min using a serial dilution of your target nucleic acid. LAMP has been used for RNA viruses (3, 4, 6, 7, 10). One-step reverse-transcription (RT)-LAMP has been achieved by initially denaturing RNA for 65°C, after which RT was done at 42°C for 1 h using *Moloney murine leukemia virus* (M-MLV) enzyme (Promega) followed by 65°C for the LAMP reaction (4). *Avian myleoblastosis virus*

(AMV) reverse transcriptase has also been used for one-step RT-LAMP with a temperature of 65°C for both (4).

12. Agar can alternatively be heated on a hot plate.

13. SYBR® Safe is regarded as having no to very low mutagenic activity (Ames tests: *see* Product Information Sheet MP 33100, Invitrogen). It is still advised to exercise appropriate care when handling the dye. SYBR® Safe dye is diluted in anhydrous dimethyl sulfoxide (DMSO) which is readily absorbed through the skin and may act as a carrier for the dye. Nitrile gloves are not recommended for use with DMSO as some brands have degradation times of 5 min. If the SYBR® Safe stain solution comes into contact with gloves, they should be changed immediately before touching equipment or work surfaces.

14. Ethidium bromide or other nucleic acid binding dyes can be used as an alternative to SYBR® Safe. Ethidium bromide is mutagenic and can be absorbed through skin and by inhalation. Always wear suitable gloves and wear a face mask when making up solutions from powder. Alternatively, ethidium bromide can be purchased as a concentrated aqueous solution (Sigma), which reduces the risk of absorption by inhalation.

15. SYBR® Safe dye is light sensitive, and excessive or prolonged exposure to light should be avoided. We recommend either casting the gel in a darkened room, or placing a box over the gel. In addition, it is recommended to run the gel in a darkened room (although gel loading can take place with the light on).

16. With HPLC-purified primers, a high-titre band of the major amplification product and several faint, larger products can be seen (Fig. 1). This banding pattern is typical when using HPLC-purified primers; use of non-HPLC-purified primers results in a lower titre of the main amplification product and more non-specific laddering.

17. For this protocol, we have described three alternative examples of end point detection methods which are straight forward to use. We have found that end point detection methods which require opening the PCR tubes following the LAMP reaction, e.g. gel electrophoresis and, PicoGreen®, increase the risk of contaminating other reaction tubes, work surfaces, and other laboratory equipments due to the large amount of target product produced during the LAMP reaction. Using a closed-tube system, such as HNB, which does not require the opening of tubes after completion of the reaction, is our preferred end point detection method due to the decreased risk of contaminating other reaction tubes, work surfaces, and equipment.

References

1. Notomi, T., Okayama, H., Masubuchi, H., Yonekawa, T., Watanabe, K., Amino, N., and Hase, T. (2000) Loop-mediated isothermal amplification of DNA. *Nucleic Acids Res.* 28, E63.

2. Nagamine, K., Hase, T. and Notomi, T. (2002) Accelerated reaction by loop-mediated isothermal amplification using loop primers. *Mol. and Cell. Probes* 16, 223–229.

3. Fukuta, S., Mizikami, Y., Ishida, A., Ueda, J., Kanabe, M. and Ishimoto,Y. (2003) detection of *Japanese yam mosaic virus* by RT-LAMP. *Arch. Virol.* 148, 1713–1720.

4. Nie, X. (2005) Reverse Transcription Loop-mediated isothermal amplification of DNA for Detection of *Potato virus Y. Plant Dis.* 89, 605–610.

5. Savan, R., Igarashi, A., Matsuoka, S. and Sakai, M. (2004) Sensitive and rapid detection of Edwardsiellosis in fish by a loop-mediated isothermal amplification method. *Appl. Environ. Microbiol.* 70, 621–624.

6. Dukes, J. P., King, D. P. and Alexandersen, S. (2006) Novel reverse-transcription loop-mediated isothermal amplification for rapid detection. *Arch. Virol.* 151, 1093–1106.

7. Varga, A. and James, D. (2006) Use of reverse transcription loop-mediated isothermal amplification for the detection of *Plum pox virus. J. Virol. Methods* 138, 184–190.

8. Tomlinson, J., Barker, I. and Boonham, N. (2007) Faster, simpler, more specific methods for improved molecular detection of *Phytophthora ramorum* in the field. *Appl. Environ. Microbiol.* 73, 4040–4047.

9. Tomlinson, J. and Boonham, N. (2008) Potential of LAMP for detection of plant pathogens. *Common W. Agric. Bur. Rev. Perspect. Agric Vet. Sci. Nutr. Nat. Resour.* 3, 1–7.

10. Fukuta, S., Ohishi, K., Yoshida, K., Mizikami, Y., Ishida, A. and Kanabe, M. (2004) Development of immunocapture reverse-transcription loop mediated isolthermal amplification for the detection of *Tomato spotted wilt virus* from chrysanthemum. *J. Virol. Methods* 121, 49–55.

11. Goto, M., Honda, E., Ogura, A., Nomoto, A. and Hanaki, K. (2009) Colormetric detection of loop mediated isothermal amplification reaction by using hydroxynapthol blue. *Biotechnqiues* 46, 167–172.

12. Iwamoto, T., Sonobe, T. and Hayashi, K. (2003) Loop-mediated isothermal amplification for direct detection of *Mycobacterium tuberculosis* complex, *M. avium*, and *M. intracellulare* in sputum samples. *J. Clin. Microbiol.* 41, 2616–2622.

13. Mori, Y., Nagamine, K., Tomita, N. and Notomi, T. (2001) Detection of loop-mediated isothermal amplification reaction by turbidity derived from magnesium pyrophosphate formation. *Biochem. Biophys. Res. Commun.* 289, 150–154.

14. Mori, Y., Hirano, T. and Notomi, T. (2006) Sequence specific visual detection of LAMP reactions by addition of cationic polymers. BMC *Biotechnol.* 6, 3.

15. Tomlinson, J. A., Dickinson, M. J. and Boonham, N. (2010) Rapid detection of *Phytophthora ramorum* and *P. kernoviae* by two-minute DNA extraction followed by isothermal amplification and amplicon detection by generic lateral flow device. *Phytopathology* 100, 143–149.

16. Harper, S. J., Ward, L. I. and Clover, G. R. G. (2010) Development of LAMP and real-time PCR methods for the rapid detection of *Xylella fastidiosa* for quarantine and field applications. *Phytopathology* 84, 456–461.

17. En, F. X., Wei, X., Jian, L. and Qin, C. (2008) Loop-mediated isothermal amplification establishment for detection of pseudorabies virus. *J. Virol. Methods* 151, 35–39.

18. Kubota, R., Vine, B. G., Alvarez, A. M. and Jenkins, D. M. (2008) Detection of *Ralstonia solanacaerum* by loop-mediated isothermal amplification. *Phytopathology* 98, 1045–1051.

19. Okuda, M., Matsumoto, M., Takana, Y., Subandiah, S. and Iwanami, T. (2005) Characterisation of the *tuf*B-*sec*E-*nus*G-*rpl*KAJL-*rpo*B gene cluster of the citrus greening organism and detection by loop-mediated isothermal amplification. *Plant Dis.* 89, 705–711.

20. Song, T., Toma, C., Nakasone, N., Iwanaga, M. (2005) Sensitive and rapid detection of *Shigella* and enteroinvasive *Escherichia coli* by a loop mediated isothermal amplification method. *FEMS Microbiol. Lett.* 243, 259–263.

Chapter 14

Genomic DNA Extraction and Barcoding of Endophytic Fungi

Patricia L. Diaz, James R. Hennell, and Nikolaus J. Sucher

Abstract

Endophytes live inter- and/or intracellularly inside healthy aboveground tissues of plants without causing disease. Endophytic fungi are found in virtually every vascular plant species examined. The origins of this symbiotic relationship between endophytes go back to the emergence of vascular plants. Endophytic fungi receive nutrition and protection from their hosts while the plants benefit from the production of fungal secondary metabolites, which enhance the host plants' resistance to herbivores, pathogens, and various abiotic stresses. Endophytic fungi have attracted increased interest as potential sources of secondary metabolites with agricultural, industrial, and medicinal use. This chapter provides detailed protocols for isolation of genomic DNA from fungal endophytes and its use in polymerase chain reaction-based amplification of the internal transcribed spacer region between the conserved flanking regions of the small and large subunit of ribosomal RNA for barcoding purposes.

Key words: Fungal endophytes, Internal transcribed spacer, Fusarium oxysporum, Polymerase chain reaction, PCR, Molecular identification techniques

1. Introduction

Fungal endophytes live inter- and/or intracellularly inside the healthy aboveground tissues of plants without causing disease (1). Endophytic fungi have been found in virtually every vascular plant species examined to date (2), and the origins of this symbiotic relationship between endophytes and plants may go back in history to the emergence of vascular plants (3). Although fungi are in fact a monophyletic sister clade to the animals (4–6), the symbiotic nature of the relationship between endophytic fungi and plants and their important contribution to the ecology of plants, justifies inclusion in this book.

Endophytic fungi receive nutrition and protection from their hosts while the plants benefit from the production of fungal secondary metabolites, which enhance the host plants' resistance to

Nikolaus J. Sucher et al. (eds.), *Plant DNA Fingerprinting and Barcoding: Methods and Protocols*, Methods in Molecular Biology, vol. 862, DOI 10.1007/978-1-61779-609-8_14, © Springer Science+Business Media, LLC 2012

herbivores, pathogens, and various abiotic stresses (7). For example, *Neotyphodium coenophialum* Glenn, Hanlin, and Bacon is an endophytic fungus found in tall fescue (*Schedonorus phoenix* Scop. Holub, formerly *Lolium arundinaceum* Schreb. S. J. Darbyshire, and formerly *Festuca arundinacea* Schreb.). Tall fescue is native to Europe but has become a major forage grass for livestock worldwide. It now covers about 14 million hectares in the USA alone because it is the only cool-season perennial forage that tolerates severe weather, inhospitable soil, and performance demands of beef operations in the southeast of the USA (8). Cattle grazing on tall fescue were found to show various symptoms of disease known as fescue foot, fat necrosis, and the general malady of fescue toxicosis (summer slump) causing economic decline due to low gains and very low calf birth weights (8). It is now known that these symptoms are caused by mycotoxins produced by *N. coenophialum*. At the same time, however, endophyte infected grass has proved to exhibit better stress and insect resistance. Intense research has led to the realization that the endophytes produce several classes of mycotoxins, of which the ergot alkaloids and indole diterpenes have been implicated in livestock toxicoses, while the loline alkaloids are far more active against insects than mammals. These observations have resulted in efforts to minimize or eliminate the ergot and indole diterpene alkaloids that are toxic to mammals but to retain the loline alkaloids, which are considered desirable because of their anti-insect activity (9). This has been achieved by the insertion of endophytes with desirable characteristics, e.g., nonergot alkaloid-producing endophytes, into noninfected tall fescue seedlings. Today, several native or novel endophyte-infected tall fescue cultivars are available that have stress-resistance mechanisms while lacking the detrimental aspects of toxicity to livestock (8).

Endophytic fungi have recently also attracted interest as potential sources of secondary metabolites with medical and industrial use (1, 10–13). The potential of endophytic fungi as sources of drugs was stimulated by the surprising discovery that certain endophytic fungi produced the anticancer drug taxol (1, 12, 13). Taxol is a diterpenoid, which was first isolated from *Taxus brevifolia*. It has a unique mode of action preventing the depolymerization of tubulin during cell division. The most common source of taxol is the bark of Yew trees. Unfortunately, however, these trees are not only rare but also slow growing, and large amounts of bark yield only small amounts of the drug, which has contributed to its high price. The discovery that endophytic fungi produce taxol has raised the possibility that this important drug may be produce via industrial fermentation.

Over the last 25 years, much progress has been made in the life history, phylogeny, and ecology of endophytic fungi (3). Yet, immense biodiversity of the kingdom fungi poses many unresolved challenges (2, 4).

Traditionally, fungal endophytes have and still are characterized morphologically, which is, however, often not sufficient to determine the species but can aid to distinguish the genus. Morphological studies aim to identify a species by its organic structure such as color, pattern, and shape. Although its appearance may benefit characterization, its results can show several species to be same, when really they are not (2).

DNA barcoding aims at reducing previous conflicting and doubtful identification analysis of several species, such as medicinal plants (14), animals (15), and fungal species (16). In an ideal future, mycologist, researchers, and amateurs around the world will be able to identify species thanks to a simple barcode, just like in the supermarkets, where each product has their own barcode. The fundamental intention of introducing DNA barcoding today aims at providing a quick and easy, yet specific method of species identification. A global pursuit of developing a standard for species identification is being lead by the Consortium for the Barcode of Life (CBoL; http://barcodeoflife.org/). CBoL promotes the benefits of using DNA barcoding in science and provides the latest developments on this automated method of species identification, which is vital information in modern taxonomy.

Ideally, just a short genetic sequence from a section of the DNA region is required (17). A typical region for eukaryotes lies within the mitochondrial DNA (mtDNA) gene known as the cytochrome c oxidase I (COI), which was first proposed by Hebert et al. (15). Although it is universally useful for characterizing animals and algae, it is not so in plants or fungi because their mitochondrial genes evolve too slowly to allow discrimination between species (2, 18). Furthermore, these genes are often duplicated in fungi (18). In fungi, the internal transcribed spacer (ITS) region between the conserved flanking regions of the small and large subunit of ribosomal RNA is the most frequently sequenced genetic marker of fungi and has been used to address research questions relating to systematics, phylogeny, and identification of strains at and even below the species level. It should be noted, however, that this region is not without potential complications and cannot serve as a single universal barcode of fungi (2). Commonly used primers for amplifying the ITS region are presented in Fig. 1 (reproduced with permission by Martin et al. (19)).

Here, we provide detailed protocols for isolation of genomic DNA from fungal endophytes and its use in polymerase chain reaction (PCR)-based amplification of ITS region between the conserved flanking regions of the small and large subunit of ribosomal RNA for barcoding purposes.

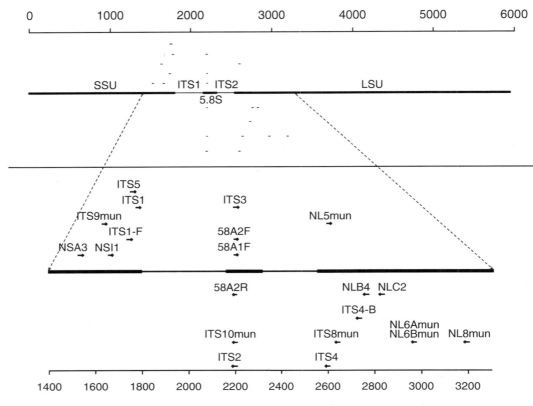

Fig. 1. An illustration of the binding positions of commonly used primers on the ribosomal cassette reproduced with permission from Martin and Rygiewicz (19). Primers are positioned above (forward primers) or below (reverse) their sequence positions. ITS1, ITS2, ITS3, and ITS4 from White et al. (20), primers ITS8mun, ITS9mun, ITS10mun, NL5mun, NL6Amun, NL6Bmun, NL8mun from Egger (22), primers ITS1-F and ITS4-B from Gardes and Bruns (23), and the remaining primers (NSA3, NSI1, 58A1F, 58A2F, 58A2R, NLB4, NLC2) from Martin et al. (23). Scale is in base pairs according to the extension of the Gargas and DePriest (24) nomenclature system described in this study.

2. Materials

2.1. Fermentation and Extraction

1. Czapek-Dox broth (Aldrich).
2. Distilled water.
3. 2×1 L Schott bottles.
4. Approximately 4×250 mL glass sterile conical flasks depending on the yield of mycelia are required (approximately 500 mg mycelia (wet weight) per 250-mL flask, depending on the species).
5. Sterile cheesecloth to cover the conical flasks (see Note 2).

2.2. Agarose Gel Electrophoresis

1. Agarose gel.
2. 50× TAE buffer.
3. Ethidium bromide (10 mg/mL).

4. 5× Nucleic acid sample loading buffer.

5. 1-kb ladder.

2.3. DNA Extraction

1. Sterile 0.22-μm vacuum filters (Millipore Express™ Plus).

2. Liquid Nitrogen.

3. Mortar and pestle.

4. DNeasy plant mini kit (QIAGEN).

2.4. PCR Amplification of Fungal ITS Barcode Regions

1. iProof High-Fidelity PCR kit (Bio-Rad).

2. DNA Engine® Peltier Thermal Cycler (Bio-Rad).

2.5. PCR Reaction Clean-Up

1. PureLink™ PCR Purification Kit (Invitrogen).

---

3. Methods

3.1. Fungus Fermentation

1. Make the liquid medium by adding 45.5 g Czapek Dox to a 1-L Schott bottle and making to volume with distilled water. Sterilize the medium and conical flasks by autoclaving for at least 15 min (see Note 1).

2. Pour the liquid medium into the 250-mL conical flasks.

3. Gently remove the fungal mycelia from the stock petri dish using a sterile disposable needle and add it to the conical flasks.

4. Cover the flasks with cheesecloth to allow aeration.

5. Incubate the flasks at 25°C with agitation at 120 rpm for 7–14 days.

3.2. Genomic DNA Extraction Using the DNeasy Plant Mini Kit

1. Filter the liquid medium containing the mycelia using a sterile 0.22-μm vacuum filter.

2. Transfer approximately 20 mg wet weight of the mycelia to a mortar and add liquid nitrogen. Grind the mycelia with a pestle under liquid nitrogen until a fine powder is produced. The liquid nitrogen will need to be constantly topped up so as to not let the mycelia thaw (see Note 3).

3. Pour the mycelia in liquid nitrogen into a 2-mL centrifuge tube and allow the liquid nitrogen to evaporate.

4. Extract the genomic DNA (gDNA) using the DNeasy® plant mini kit, as described by the manufacturer (Qiagen DNeasy® Plant Handbook, 07/2006, p:2. step 7). Use distilled water as the final elution solvent instead of buffer AE.

3.3. Assessment of DNA Quantity and Purity by UV Spectrophotometry

1. Measure the UV spectrum of the DNA solution between 220 and 320 nm. The absorbance should be between 0.1 and 1 AU.

2. Estimate the concentration of DNA by measuring the absorbance of the solution at 260 nm. Multiply the absorbance by

50 μg/mL/AU to obtain an approximate concentration of dsDNA.

3. Estimate the purity of the DNA from proteins by measuring the ratio between the absorbance at 260 nm and the absorbance at 280 nm. Pure DNA has an A_{260}/A_{280} ratio of 1.7–1.9.

4. Estimate the purity of the DNA from EDTA, carbohydrates and phenol by measuring the ratio between the absorbance at 260 nm and the absorbance at 230 nm. Pure DNA has an A_{260}/A_{230} ratio of 2.0–2.3.

3.4. Assessment of gDNA Quality by Agarose Gel Electrophoresis

1. Prepare a 0.7% agarose gel by dissolving 0.35 g of agarose powder in 50 mL of 1× TAE.

2. Load the gDNA (at least 20 ng) mixed with loading buffer onto the gel. Use a 1-kb ladder for size reference.

3. Run the gel at 100 V for approximately 60 min.

4. Stain the gel in ethidium bromide standing solution (0.5 μg/mL).

5. Visualize the gel under UV light to determine gDNA quality.

A single band should be evident at the top of the gel. If the gDNA has been damaged, then streaking usually occurs. Streaking is not necessarily a problem for DNA barcoding as there will almost definitely be enough of the template gene to replicate by PCR.

3.5. PCR Amplification of Fungal ITS Barcoding Regions

1. Setup the PCR reactions using the iProof hi-fidelity DNA polymerase kit according to the manufacturer's instructions. We generally use 50-uL reaction per 1 unit of iProof DNA polymerase to ensure enough DNA is produced for purification, quality control and sequencing. The primer sequences are presented in Table 1 and the reaction volumes are presented in Table 2.

2. Amplify the template DNA using the following temperature program adapted from White et al. (20): initial denaturation 95°C,

Table 1
A list of commonly used primers for barcoding fungi

Primers	Sequence 5′–3′
NSI1 (forward)	GAT TGA ATG GCT TAG TGA GG
NLB4 (reverse)	GGA TTC TCA CCC TCT ATG AC
ITS1 (forward)	TCC GTA GGT GAA CCT GCG G
ITS4 (reverse)	TCC TCC GCT TAT TGA TAT GC

Table 2
Reaction volumes for PCR amplification

Component	Volume per reaction
5× Proof buffer	10 μL
dNTP mix	1 μL
Forward primer (10 μM)	2.5 μL
Reverse primer (10 μM)	2.5 μL
DNA template	50 ng
Proof DNA polymerase	0.5 μL
Sterile H$_2$O	Make to 50 μL

60 s; denaturation 95°C, 30 s; annealing 60°C, 40 s; extension 72°C, 90 s; final extension time 72°C, 300 s. Denaturation, annealing, and extension were repeated 35 times.

3.6. Assessment of PCR Product Quality by Agarose Gel Electrophoresis

1. Prepare a 1.0% agarose gel by dissolving 0.5 g of agarose powder, in 50 mL of 1× TAE.

2. Load at least 20 ng of the PCR product mixed with loading buffer onto the gel. Use a 1-kb ladder for size reference.

3. Run the gel at 100 V for approximately 60 min.

4. Stain the gel in ethidium bromide standing solution (0.5 μg/mL).

5. Visualize the gel under UV light to determine gDNA quality.

A single band should be evident. If there are multiple bands or no bands then the general PCR conditions given above will need to be modified. Generally, the annealing temperature and Mg^{2+} concentration are useful parameters to adjust.

3.7. PCR Clean-Up

1. Clean the PCR amplification product using the PureLink™ PCR Purification Kit, as described by the manufacturer's instructions, except for the following changes:

    (a) Buffer HC was used for DNA binding to remove low molecular weight PCR by-products.

    (b) 30 μL of distilled water was used to elute the purified DNA.

2. Assess the purity and quantity of the amplicon as described in Subheadings 3.3 and 3.6.

4. Notes

1. Sterile cheesecloth or gauze is required to minimize contamination by airborne fungus while allowing air exchange.

2. Fungal cultures can be easily contaminated. It is extremely important that all glasswares, tools, and work areas are sterile.

3. Generally speaking, grinding the mycelia with a mortar and pestle under liquid nitrogen gives better quality gDNA in greater yield than using the TissueLyser procedure recommended by QIAGEN.

Acknowledgments

We thank Dr Jeremy Burdon and Dr Bo Wang from the CSIRO Plant Industry in Canberra for providing us with their extensive collection of endophytes isolated from wild cotton species in Australia (21). Many thanks to Dr Kendall Martin and Dr Paul Rygiewicz for granting use of Fig. 1.

References

1. Tan RX and Zou WX (2001) Endophytes: a rich source of functional metabolites. Natural Product Reports 18(4):448–459

2. Rodriguez RJ, White JF, Arnold AE, et al (2009) Fungal endophytes: diversity and functional roles. New Phytologist 182(2): 314–330

3. Zhang HW, Song YC, and Tan RX (2006) Biology and chemistry of endophytes. Nat Prod Rep 23(5):753–771

4. Seifert KA (2009) Progress towards DNA barcoding of fungi. Molecular Ecology Resources 9:83–89

5. McLaughlin DJ, Hibbett DS, Lutzoni F, et al (2009) The search for the fungal tree of life. Trends Microbiol 17(11):488–497

6. Stajich JE, Berbee ML, Blackwell M, et al (2009) The fungi. Curr Biol 19(18): R840–845

7. Saikkonen K, Faeth SH, Helander M, et al (1998) Fungal endophytes: A continuum of interactions with host plants. Annual Review of Ecology and Systematics 29:319–343

8. Belesky DP and Bacon CW (2009) Tall fescue and associated mutualistic toxic fungal endophytes in agroecosystems. Toxin Reviews 28(2–3):102–117

9. Schardl CL, Grossman RB, Nagabhyru P, et al (2007) Loline alkaloids: Currencies of mutualism. Phytochemistry 68(7):980–996

10. Schulz B, Boyle C, Draeger S, et al (2002) Endophytic fungi: a source of novel biologically active secondary metabolites. Mycological Research 106:996–1004

11. Strobel G (2006) Harnessing endophytes for industrial microbiology. Current opinion in microbiology 9(3):240–244

12. Strobel G and Daisy B (2003) Bioprospecting for microbial endophytes and their natural products. Microbiol Mol Biol Rev 67(4):491–502

13. Strobel G, Daisy B, Castillo U, et al (2004) Natural products from endophytic microorganisms. Journal of natural products 67(2):257–268

14. Sucher NJ and Carles MC (2008) Genome-based approaches to the authentication of medicinal plants. Planta Med 74(6):603–623

15. Hebert PD, Cywinska A, Ball SL, et al (2003) Biological identifications through DNA barcodes. Proc Biol Sci 270(1512):313–321

16. Stockinger H, Kruger M, and Schussler A (2010) DNA barcoding of arbuscular mycorrhizal fungi. New Phytol 187(2):461–474

17. Hebert PD and Gregory TR (2005) The promise of DNA barcoding for taxonomy. Syst Biol 54(5):852–859

18. Chase MW and Fay MF (2009) Barcoding of Plants and Fungi. Science 325(5941):682–683

19. Martin KJ and Rygiewicz PT (2005) Fungal-specific PCR primers developed for analysis of the ITS region of environmental DNA extracts. BMC Microbiol 5:28

20. White T, Bruns T, Lee S, et al (1990) Amplification and direct sequencing of fungal ribosomal RNA genes for phylogenetics. In: Innis M, Gelfand D, Sninsky J, & White T (ed) PCR Protocols: A Guide to Methods and Applications. Academic Press Inc, New York

21. Wang B, Priest MJ, Davidson A, et al (2007) Fungal endophytes of native Gossypium species in Australia. Mycol Res 111(3):347–354

22. Egger KN (1995) Molecular Analysis of Ectomycorrhizal Fungal Communities. Can J Bot 73:S1415–S1422

23. Gardes M and Bruns TD (1993) Its Primers with Enhanced Specificity for Basidiomycetes – Application to the Identification of Mycorrhizae and Rusts. Mol Ecol 2(2): 113–118

24. Gargas A and DePriest PT (1996) A nomenclature for fungal PCR primers with examples from intron-containing SSU rDNA. Mycologia 88(5):745–748

Chapter 15

Using GenBank® for Genomic Authentication: A Tutorial

James R. Hennell, Paul M. D'Agostino, Samiuela Lee, Cheang S. Khoo, and Nikolaus J. Sucher

Abstract

The GenBank® database is perhaps one of the most important repositories of genetic information. A researcher working in the field of genomic authentication must therefore be equipped with the skills needed to competently access the required information from this database whilst ultimately contributing their own data to it. This chapter presents a practical guide to using GenBank® to search for sequences, search and align unknown sequences using BLAST, and uploading and maintaining your own sequences. This chapter also details some other software helpful in sequence manipulation.

Key words: GenBank, BLAST, Sequin, BankIt, Alignment, Sequencing

1. Introduction

To take advantage of the enormous amount of biological data that is being continually generated, the data needs to be managed systematically so that it is conveniently accessible and searchable (1, 2). This is especially true with respect to genetic information, where the large repository GenBank® doubles in size every 18 months and now contains over 108 million entries for over 260,000 named organisms (3). To use this resource effectively, it is necessary to have knowledge of how all the desired data can be extracted while filtering out extraneous information (2).

This chapter describes the common techniques employed by molecular biologists using the GenBank® database (see Note 1). Before introducing the techniques for using the database, a description of the resource and terminologies used are given.

1.1. Description of GenBank®

GenBank® is an online genetic sequence database built by the National Center for Biotechnology Information (NCBI), a division of the National Library of Medicine (NLM) on the National

Nikolaus J. Sucher et al. (eds.), *Plant DNA Fingerprinting and Barcoding: Methods and Protocols*, Methods in Molecular Biology, vol. 862, DOI 10.1007/978-1-61779-609-8_15, © Springer Science+Business Media, LLC 2012

Institutes of Health (NIH) campus (3). GenBank® collaborates with organisations such as the European Molecular Biology Laboratory Nucleotide Sequence Database (EMBL) and the DNA Databank of Japan (DDBJ) to form the International Nucleotide Sequence Database Collaboration (INSDC), though each database exists as a separate entity. Data is uploaded into each database separately using their own tools but the data is synchronised between each database daily to ensure each contains current information. This chapter only focuses on exploring data contained in the GenBank® database.

A key tool used is the Entrez cross-database search system, another initiative of the NCBI. Entrez is a unified search platform used to interface simultaneously with many of the databases contained at the NCBI such as the nucleotide, protein, genome and taxonomy databases. This information is relevant because it can be used to access the nucleotide database (at GenBank®) as well as the genome database that can be very useful for identifying fully genome-sequenced organisms. The Entrez page can be accessed at http://www.ncbi.nlm.nih.gov/sites/gquery.

1.2. FASTA Format

In general, FASTA refers to the FASTA data formatting style used to encode a single line sequence definition, sequence information and alignment information into one file that can be accessed by GenBank® and many other programs.

The definition line starts with a ">" symbol followed by a unique sequence identifier (SeqID) and modifiers such as the organism name in the general format (modifier = text). A list of useful modifiers is presented in Table 1.

The subsequent lines are used for the genetic sequence, where each nucleotide is represented by its single-letter IUPAC (International Union of Pure and Applied Chemistry) code with not more than 80 characters per line. For completeness, a list of the IUPAC nucleotide abbreviations is listed in Table 2.

1.3. Searching GenBank®

The GenBank® Web site can be accessed by selecting the "nucleotide" link at the Entrez homepage found at http://www.ncbi.nlm.nih.gov/sites/gquery or by the direct link http://www.ncbi.nlm.nih.gov/nuccore. A basic search, such as looking up records based on keywords can be performed using the search window. More complex searches, such as searching for a sequence match or alignment are accessed through the Basic Local Alignment Search Tool (BLAST) link.

The BLAST is used to compare an unknown DNA sequence rapidly and accurately to known and well-characterised sequences contained in the GenBank® database and is one of the most important techniques used in plant genomic authentication (4). While BLAST can also be used for protein searching, only nucleotide searching is covered in the scope of this chapter.

Table 1
A list of valid modifiers for the FASTA definition line (18)

Modifier

Acronym	Forma	rev-PCR-primer-name
anamorph	forma-specialis	rev-PCR-primer-seq
authority	fwd-PCR-primer-name	segment
bio-material	fwd-PCR-primer-seq	serogroup
biotype	genotype	serotype
biovar	group	serovar
breed	haplogroup	sex
cell-line	haplotype	specimen-voucher
cell-type	host	strain
chemovar	identified-by	sub-species
chromosome	isolate	subclone
Clone	isolation-source	subgroup
clone-lib	lab-host	substrain
collected-by	lat-lon	subtype
collection-date	linkage-group	synonym
common	map	teleomorph
country	note	tissue-lib
cultivar	organism	tissue-type
culture-collection	pathovar	type
dev-stage	plasmid-name	variety
ecotype	plastid-name	
endogenous-virus-name	pop-variant	

The strength of the BLAST program is in its heuristic nature, where small assumptions are made in order to speed up (though with less accuracy) the search, where using a more exhaustive search method such as the FASTA sequence comparison program would be impractically slow (5).

1.4. Submitting Your Sequence to GenBank®

It is becoming common practice amongst molecular biological journals to require the submission of sequence data to a publicly available database such as GenBank® before publication. Sequences are generally submitted to GenBank® electronically by the researcher using either New BankIt or Sequin (3).

Table 2
IUPAC nucleotide abbreviations (19)

Nucleotide	Abbreviation
Adenosine	A
Cytidine	C
Guanine	G
Thymidine	T
Uridine	U
G or A	R
T or C	Y
G or T	K
A or C	M
G or C	S
A or T	W
G or T or C	B
G or A or T	D
A or C or T	H
G or C or A	V
A or G or C or T	N
Gap of indeterminate length	-

1.5. Analysis and Manipulation of Sequencing Data

When comparing a sequence from an unknown plant against a GenBank® record for genomic authentication, it is important that the unknown sequence is high quality. Errors of only a few nucleotides can have the potential to give varying results depending on the organism.

Sequencing data is usually provided in two formats: (1) a FASTA file of the raw sequence and (2) a chromatogram file (also known as a trace file) of the dye-terminator labelled DNA. Though it is possible to perform a nucleotide BLAST from the raw FASTA file sequence as a quick check, it is necessary to analyse the chromatogram output from the sequencer to determine the quality of the sequence, as there is often lower quality data at the beginning and end of a sequence.

2. Program Usage

2.1. The BLAST Program

The BLAST program is initiated by sending the query sequence as well as any specific details (such as word size) to the BLAST server found at http://blast.ncbi.nlm.nih.gov/Blast.cgi. The program works by firstly breaking the query sequence into several shorter sequences known as "words". Segment pairs are then created by aligning each of the query words with each of the database sequences that are also broken into words. These segment pairs are then given a bit score based on their similarity. The program then finds high-scoring segment pairs (HSP), which have a bit score exceeding a particular cut-off value and extends the length of the word (with and without gaps) to create the final alignment with a maximum bit score. After the program has created these alignments between all query and database words, it assembles the best alignment for each query and database sequence pair (6, 7).

Word size determines the minimum number of bases that must match between the query and database sequence. Increasing the word size will decrease the time taken for a query and identify database sequences that match the query very closely. Decreasing the word size will increase the sensitivity and the time taken for the query.

An indication of the quality of the BLAST result is given using the expect value and the bit score. The expect value (E-value) is an estimation of the statistical significance of an alignment, which is whether the result could be attributed to chance alone or whether it is likely a real match. The lower the E-value, the less likely that the result is due to chance (6, 8). The bit score gives the quality of the alignment, taking into account the number of matched letters, mismatched letters and gaps. The bit score gives an indication of how good the alignment is; the higher the score, the better the alignment (6, 8).

2.2. GenBank® Nucleotide Databases Searched by BLAST

NCBI contains a number of databases that can be selected to more specifically tailor the results of the BLAST query. The default database is the Human genome and should be changed to one of the suggested databases presented in Table 3 to be useful for plant genomic authentication. Generally, the nr/nt database should be used for the most comprehensive search.

2.3. Selecting the Nucleotide BLAST Program to Use

There are three BLAST programs pertinent to the genetic authentication of plant material, each tailored to search the GenBank® database in slightly different ways. A summary when to use the different BLAST programs is detailed below.

2.3.1. MEGABLAST

MEGABLAST is a program tailored to rapidly align long query sequences to very similar database sequences (9). MEGABLAST is

Table 3
A suggested list of useful GenBank® databases for plant genomic authentication (4)

Database	Description
nr/nt	This includes the entire GenBank® nucleotide collection and should be used to get the most comprehensive results
Htgs	Unfinished High Throughput Genomic Sequences (phases 0, 1 and 2). Finished sequences (phase 3) are in the nr/nt database
Month	All new or revised GenBank sequences released in the last 30 days
chromosome	Complete genomes and complete chromosomes from the NCBI Reference Sequence project
Wgs	Assemblies of Whole Genome Shotgun sequences
env_nt	Sequences from environmental samples, such as uncultured bacterial samples isolated from soil or marine samples. This does not overlap with the nr/nt database

intended for comparing a query to a database sequence with a match of >95%. MEGABLAST is therefore recommended for genomic authentication, as there should be an exact match to the submitted query sequence in the database.

2.3.2. Discontiguous MEGABLAST

Discontiguous MEGABLAST is used to find similar alignments to related database sequences from other organisms rather than to find an exact match to a database sequence. Discontiguous MEGABLAST is generally considered more sensitive and efficient than standard Blastn when using the same word size and has largely replaced Blastn (4).

2.3.3. Blastn

One of the major benefits of Blastn is that it utilises smaller word sizes so it is a generally more sensitive query technique than MEGABLAST, though it has generally now been replaced by Discontiguous MEGABLAST (4).

2.4. Aligning Two or More Known Sequences

The use of BLAST for similarity searching and alignment of query and database sequences has been discussed, but if the sequences to be aligned is already known, they can be fed directly into the BLAST program (bl2seq) by selecting the "align two or more sequences" box. The techniques to perform the BLAST alignment are the same as those for an unknown sequence.

2.5. Submitting Sequence Data to GenBank®

There are two main ways to submit genetic information: using BankIt and Sequin. After the sequence has been submitted, the person submitting will be given an accession number from the database that can be used by the individual in a publication to refer to the sequence.

2.4.1. BankIt

New BankIt is a Web-based submission tool whereby data is submitted through an online form using a Web browser (10). New BankIt is the latest version of the BankIt submission process, superseding the previous version called Old BankIt.

To use BankIt for sequence submission, it is necessary to create an account on the Primary Data Archive (PDA) Login System at https://pdalogin.ncbi.nlm.nih.gov/. Your contact information is stored on the PDA to expedite future BankIt submissions. Partially completed submissions can be saved on the PDA and can be completed at a later date.

2.4.2. Sequin

Sequin is a computer program that can be downloaded to a computer and used to prepare a submission without the need for an Internet connection (11, 12). Sequin is a more powerful tool than BankIt and can automatically perform many tasks, making Sequin more useful for complex or multiple submissions. The submission is saved locally on the computer as a *.sqn file and can be edited anytime. Once the submission is complete, this file can be emailed to the NCBI. The Sequin program can be downloaded from http://www.ncbi.nlm.nih.gov/Sequin/.

By default, Sequin is not network aware (i.e. it cannot connect to the Internet). Sequin can be made network aware by selecting "network configure" on the Sequin welcome screen and setting up the Internet connection. Network aware Sequin can also be used to download an existing sequence from Entrez.

2.4.3. UpdateMacroSend

The UpdateMacroSend tool can be used as an alternative for emailing submissions or updates to NCBI (13) and is generally used to send files too large for email. It can be found at the following URL: http://www.ncbi.nlm.nih.gov/projects/GenBankUpdate/gen-bank_update.cgi.

2.6. Updating GenBank® Records

Updates to a GenBank® entry after submission can be emailed directly to the NCBI at admin@ncbi.nlm.nih.gov, sent using UpdateMacroSend or updated through Sequin. Updates may be necessary if, after the publication, details become available, or if it is necessary to annotate the sequence. To make sure the update can be processed expediently, there are strict formatting requirements, detailed in the examples section.

2.7. Applied Biosystems™ Sequence Scanner

Sequence scanner is a free program available from the Applied Biosystems™ Web site http://www.appliedbiosystems.com/absite/us/en/home/support/software.html and can be used to visually analyse trace files (extension *.abl).

3. Examples

3.1. Viewing and Manipulating Sequences Using Applied Biosystems™ Sequence Scanner

Trace files with the extension *.ab1 can be imported into the Sequence Scanner and viewed using the "Analysed" tab. The most of important features of the trace are the sequence, signal intensity, trimming bar and the quality bars as illustrated in Fig. 1. To prepare a sequence for genetic authentication via BLAST, the sequence should be edited following these steps:

1. Excise low quality bases from the terminal ends of the trace file using the trimming bar. Trimmed sequences will now appear grey.

2. Ensure the signal is clean throughout the rest of the sequence. Poor quality bases can arise from sequence contamination or low signal intensity.

3. If the sequence is high quality, it can be exported using the export function. Ensure that "post-trimmed sequence" is checked before exporting the file.

4. Import the FASTA file or copy the sequence text into GenBank® and follow the steps outlined in Subheading 3.

3.2. Initiating a BLAST Search

The nucleotide BLAST page can be accessed at http://blast.ncbi.nlm.nih.gov/Blast.cgi and selecting the nucleotide blast option as presented in Fig. 2. At the Blastn suite page, the sequence can be entered or a FASTA file uploaded to BLAST. The database from the Human genomic + transcript database (default) must be changed to the Nucleotide collection (nr/nt) or another suitable database.

The procedure is as described:

1. Enter the query sequence or upload a FASTA file.

2. Select a job title (optional).

3. Change the query database from the Human genomic + transcript database to the Nucleotide collection (nr/nt) or suitable database.

4. Select the BLAST program to use (MEGABLAST is the default).

5. Adjust the program as necessary. Generally word size is a useful parameter to adjust.

6. Start the BLAST.

3.3. Understanding the Results from the BLAST Query

This tutorial follows on from Subheading 3.2 detailing the result from the BLAST query. The results page contains a large amount of information, most of which is explained here. Additional explanations may be obtained from the NCBI learning centre available at http://www.ncbi.nlm.nih.gov/staff/tao/URLAPI/ and selecting "new blast output format explained".

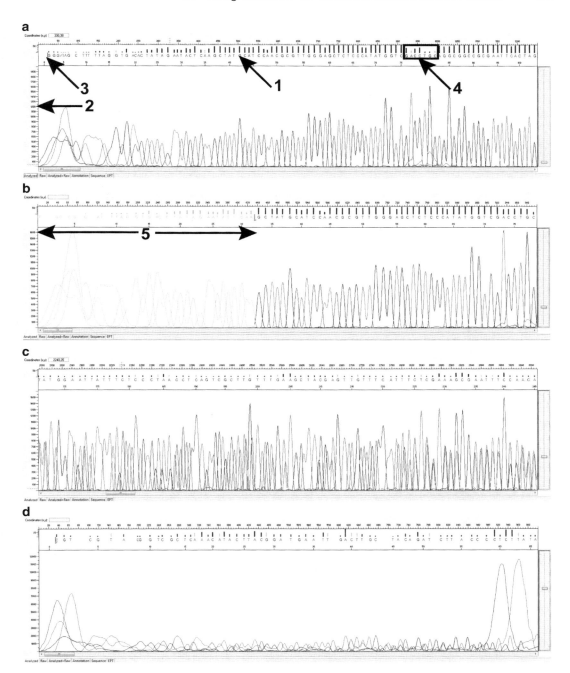

Fig. 1. Screenshots from the Sequence Scanner used to manipulate sequence data. (**a**) The "analysed" window used to display the nucleotide trace; (1) the nucleotide sequence; (2) signal intensity; (3) the trimming bar; (4) the nucleotide quality bar. (**b**) The trimming bar has been moved to disregard low quality sequences. (5) The trimmed regions are shown in *grey* and will be excluded when the nucleotide sequence is exported. (**c**) An example of a poor quality sequence where overlapping nucleotide signals and poor quality bars indicate contamination. (**d**) An example of a poor quality sequence due to poor signal intensity.

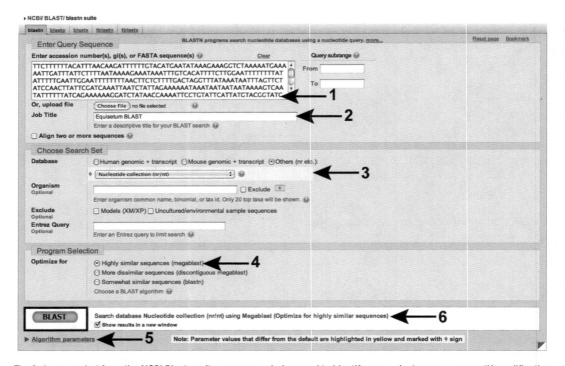

Fig. 2. A screenshot from the NCBI Blastn suite query page being used to identify an *equisetum arvense matK* amplification product. (1) Enter the query sequence; (2) Select a job title; (3) Select the database to search; (4) Select the BLAST algorithm to use; (5) Adjust the algorithm as necessary; (6) Start the BLAST.

Fig. 3. The BLAST header.

3.3.1. Header

The header section contains general information about the query parameters as detailed in Fig. 3. The following sections giving a graphical summary, descriptions and alignments go into greater detail.

3.3.2. Graphical Summary

The top returns from the alignment are presented in the graphical summary section of the BLAST report as shown in Fig. 4. The query sequence is the red line at the top of the figure, with the bases numbered. The database sequence alignments, listed in

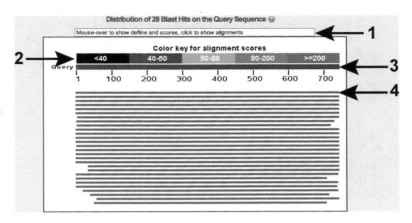

Fig. 4. The BLAST graphic summary: (1) The window that displays the definition line of the sequence the mouse is over; (2) The alignment bit score colour score key; (3) The query sequence; (4) The database sequences aligned to the query sequence. Sequences are coloured according to the key and are in order of their alignment bit score.

order of similarity, are illustrated by the thinner bars below the red query bar. The degree of similarity between the query sequence and the database sequences is colour coded, with black being little relationship and red being high alignment, as determined by the bit score. In this case, all hits have a high degree of similarity. Placing the cursor over the query bars displays the definition line for that sequence in the window at the top of the graphic and clicking the query bar scrolls down the results page to display that alignment in detail (see Note 2).

3.3.3. Descriptions

Detailed information on the database matches is provided in the descriptions section, detailed in Fig. 5 (14). The table has eight columns:

1. Accession: The unique identifier for the database sequence.
2. Description: A description of the sequence from the definition line including the organism and the gene the sequence is from.
3. Max Score: The highest bit score of the match between the query and database sequence.
4. Total Score: The sum of the score of all HSPs from the same database sequence.
5. Query Coverage: The length of the query sequence covered by different high-scoring pairs from the same database sequence.
6. E-value: The "expect value" is an estimate of statistical significance, which is basically the number of database matches that could occur due to chance. The smaller the value, the less likely that the database match is due to chance (15).

1	2	3	4	5	6	7	8

Sequences producing significant alignments:

Accession	Description	Max score	Total score	Query coverage	E value	Max ident	Links
AM883547.1	Equisetum arvense plastid partial matK gene for maturase	1365	1365	100%	0.0	99%	
AM883546.1	Equisetum arvense plastid partial matK gene for maturase	1354	1354	100%	0.0	99%	
GU191334.1	Equisetum arvense chloroplast, complete genome	1349	1349	100%	0.0	99%	
AY348551.1	Equisetum arvense tRNA-Lys (trnK) gene, intron; and MatK	1349	1349	100%	0.0	99%	
AM883545.1	Equisetum x dycei plastid partial matK gene for maturase	1317	1317	100%	0.0	98%	
AM883541.1	Equisetum palustre plastid partial matK gene for maturase	1177	1177	100%	0.0	95%	
AM883536.1	Equisetum palustre plastid partial matK gene for maturase	1170	1170	99%	0.0	95%	
AM883539.1	Equisetum palustre plastid partial matK gene for maturase	1146	1146	97%	0.0	95%	
AM883538.1	Equisetum sylvaticum plastid partial matK gene for matura	1120	1120	96%	0.0	94%	
AM883555.1	Equisetum x fontqueri plastid partial matK gene for matura	1105	1105	100%	0.0	93%	
AM883553.1	Equisetum sylvaticum plastid partial matK gene for matura	1105	1105	100%	0.0	93%	
AM883540.1	Equisetum telmateia plastid partial matK gene for maturas	1105	1105	100%	0.0	93%	
AM883552.1	Equisetum scirpoides plastid partial matK gene for matura	965	965	100%	0.0	90%	
AM883534.1	Equisetum scirpoides plastid partial matK gene for matura	963	963	99%	0.0	90%	
AM883549.1	Equisetum giganteum plastid partial matK gene for matura	920	920	100%	0.0	89%	
AM883551.1	Equisetum ramosissimum plastid partial matK gene for ma	894	894	100%	0.0	88%	
AM883543.1	Equisetum hyemale plastid partial matK gene for maturase	894	894	100%	0.0	88%	
AM883542.1	Equisetum hyemale plastid partial matK gene for maturase	894	894	100%	0.0	88%	
AM883550.1	Equisetum myriochaetum plastid partial matK gene for ma	857	857	95%	0.0	88%	
AM883548.1	Equisetum giganteum plastid partial matK gene for matura	857	857	95%	0.0	88%	
AM883544.1	Equisetum ramosissimum subsp. debile plastid partial mat	856	856	99%	0.0	87%	
EU749487.1	Equisetum hyemale voucher OAC:JAG390 maturase K (mat	839	839	95%	0.0	88%	
AM883554.1	Equisetum variegatum plastid partial matK gene for matur	833	833	99%	0.0	87%	
AM883537.1	Equisetum variegatum plastid partial matK gene for matur	828	828	99%	0.0	87%	
AM883535.1	Equisetum variegatum plastid partial matK gene for matur	797	797	96%	0.0	86%	
EU749485.1	Equisetum hyemale voucher OAC:JAG303 maturase K (mat	791	791	90%	0.0	87%	
EU749484.1	Equisetum hyemale voucher OAC:AP344 maturase K (matk	782	782	92%	0.0	87%	
EU749486.1	Equisetum hyemale voucher OAC:JAG356 maturase K (mat	771	771	88%	0.0	87%	

Fig. 5. An example of the BLAST descriptions page.

7. Max ident: The maximal identity is the number of identical nucleotides between the query sequence and the database sequence.

8. If the sequence contains records in another database, the link to that database will appear.

3.3.4. Alignments

The actual pair-wise alignments between the query sequence and each database sequence is presented in the alignments section. The alignment starts with the full definition line of the database sequence, as well as most of the data contained in the descriptions section as well as the number of gaps in the alignment.

A representative sequence alignment is presented in Fig. 6. Important information to note about the alignment is:

- The query sequence is on top, the database sequence is at the bottom.

- The nucleotide number is given on the left and right of the alignment.

- The vertical line between the sequences indicates a match. A mismatch will have no line.

- A dash indicates a gap.

- Lowercase letters indicate letters that are not used in the query alignment because they are repeated or of low complexity.

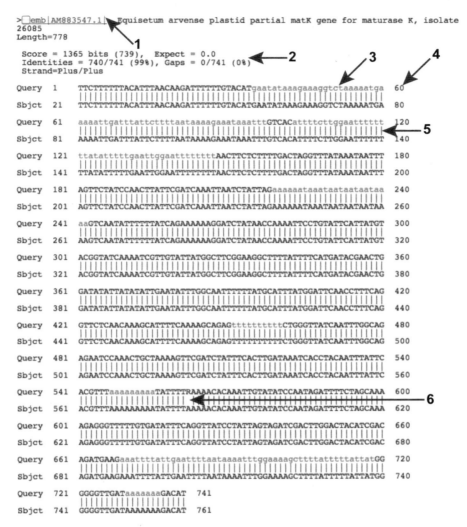

Fig. 6. An overview of an alignment result: (1) The definition line of the database sequence being aligned; (2) A detailed summary of BLAST statistics; (3) Lowercase letters not used in the BLAST; (4) The nucleotide number; (5) Vertical lines representing a match; (6) A gap representing a mismatch.

3.4. Aligning Two or More Known Sequences

If the sequences to be aligned are known, they can be directly selected for alignment at the BLAST homepage as described in Fig. 7. In this example, the *Equisetum arvense matK* sequence used above is being aligned with a GenBank® record that contains the complete *E. arvense* chloroplast genome (accession GU191334). The results from the alignment are in the same format as the results from Subheading 3.3.

3.5. Updating GenBank® Entries

A list of modifications that can be made once a sequence has been submitted to GenBank® are as follows (16). The given example is for modifications that may be made to the *E. arvense matK* nucleotide

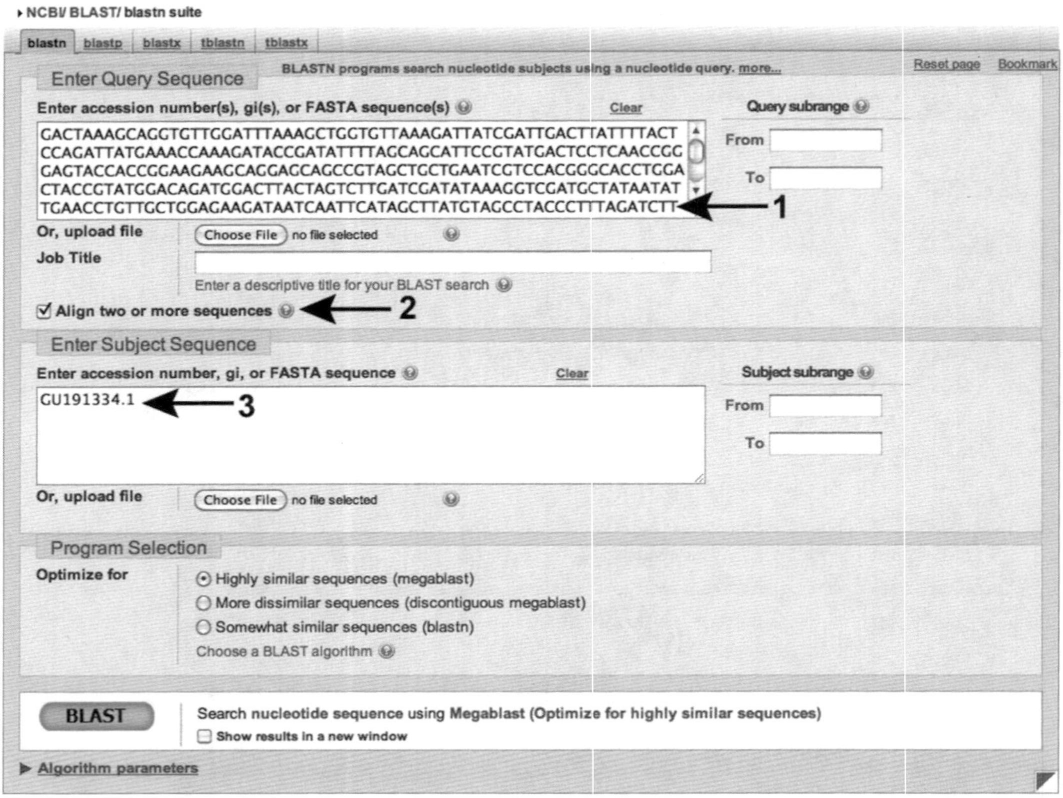

Fig. 7. A screenshot from the BLAST homepage showing the "align two or more sequences" option: (1) The query sequence; (2) The "align two or more sequences" option; (3) The sequence(s) you want to align with the query sequence.

sequence after it has been released, with an invented accession number AYxxxx02. These modifications can generally be emailed to the NCBI (admin@ncbi.nlm.nih.gov), where a technician will then make the changes to the database.

3.5.1. Source Information

Any additional modifiers for the definition line can be added in a multi-column tab-delimited table as follows. In this example, some primer information and the country of origin are added to the record.

acc. num.	rev-PCR-primer-name	rev-PCR-primer-seq	country
AYxxxx02	matK F (equisetum)	ATACCCCATT TTATTCATCC	Germany

3.5.2. Publication Information

Any additional publication information can be annotated to the sequence, such as additional authors or publication details by emailing the GenBank accession number(s) and all relevant publication information. The author lists should be formatted as `Last_Name(family name), First_Initial, Middle_ Initial`. The complete journal name should be included (not the abbreviation).

3.5.3. Nucleotide Sequence

The entire nucleotide sequence can be updated by emailing the new nucleotide sequence in FASTA format. The NCBI will not accept a list of nucleotide changes or non-IUPAC characters.

3.5.4. Update Features on Record Without Annotation

New annotations can be added to a nucleotide record by emailing the features using either a tab-delimited five-column feature table or a spreadsheet of the features. An example of adding a feature to the *E. arvense matK* sequence is shown below; sections changed are given in Fig. 8.

Tab-Delimited Five-Column Feature Table

The feature table specifies the location and type of each feature with the columns separated by tabs. The first line of the table contains the sequence identifier.

`>Feature SeqId`

The other lines contain the feature information in the following format:

Line 1: Information on the feature type and location.

Tab 1: The start nucleotide of a feature.

Tab 2: The stop nucleotide of a feature.

Tab 3: The name of the feature such as CDS or gene.

```
AUTHORS    Hennell,J.R., Lee,S., Cook,R., Carles,M.C., Low,M.N., Lamin,F.,
           Khoo,C.S., Higgins,V.J., Govindaraghavan,S. and Sucher,N.J.
TITLE      Direct Submission
JOURNAL    Submitted (20-MAR-2011) Centre for Complementary Medicine Research,
           University of Western Sydney, Locked bag 1797, Penrith, NSW 2750,
           Australia
FEATURES             Location/Qualifiers
     source          1..741
                     /organism="Equisetum arvense"
                     /organelle="plastid:chloroplast"
                     /mol_type="genomic DNA"
     gene            <1..>741
                     /gene="matK"
     CDS             <1..>741
                     /gene="matK"
                     /codon_start=2
                     /transl_table=11
                     /product="maturase K"
                     /translation="SFLHLTRFFVHEYKERSKNEKLIYSFNKRNKFVTFSWNFFYIFE
                     LEFFLTSLLTRFINNLVLSNLFDQINLLEKINNNKSQYFLSEKRIYNQNSCIHYVRY
                     QNRCIMASEGFYFHDTNWIYYILNIWQFFMHLWIQPFRFSTKHFQKQSFFFLGYQFGR
                     ESKLLKVRSISLDKSPTIYSRLKKNILKTQIVYPIDFLAKEGFCDISGYPISRSTWTT
                     STDEEILLNFNKIWKSFYFYYGGLIKKD"
BASE COUNT      265 a      83 c      97 g     295 t        1 others
```

Fig. 8. Details in the "features" section are coloured to highlight what the different feature table or spreadsheet entries change. *Grey* changes the gene entry, *black* changes the CDS entry.

Line 2: Qualifiers to further detail the feature.

Tab 4: The name of the qualifier such as gene or product.

Tab 5: The value of the qualifier.

Further features can be added in the above format. An example of a correctly formatted feature table is presented below, where a gene and CDS annotation are added to the *matK* sequence.

```
>Feature AYxxxx02

<1    >741 gene

                    gene matK

<1    >741 CDS

                    product    matK

                    codon_start    2
```

Spreadsheet

The annotation can also be emailed as a spreadsheet, which may be more user-friendly to read as presented in Table 4. The standard format for the spreadsheet is column 1 which is the accession number, column 2 is the feature type (e.g. CDS) and column 3 is nucleotide location (given as a range) of the feature. After the first three columns, feature qualifiers may be listed in any order (17). Again the start and/or stop locations should be prefixed with a < and >, respectively, if the feature is partial.

3.5.5. Update Features on Record with Annotation

Updates to the annotation of a sequence that already has annotations can be emailed to the NCBI with the accession number of the sequence. The NCBI will email the current annotation of the sequence in a tab-delimited five-column feature table that can then be updated with the new data and emailed back.

3.5.6. Network Aware Sequin

The network aware Sequin can be used to download an existing sequence from Entrez for changes to be made. Network aware Sequin should not be used for simple updates such as publication

Table 4
An example of a correctly formatted feature spreadsheet

Accession	Feature	Location	Product	codon_start	Gene
equi_europe	gene	<1..>741			matK
	CDS	<1..>741	matK	2	

changes or source information changes. The updated *.sqn file should then be emailed to gb-admin@ncbi.nlm.nih.gov or sent using UpdateMacroSend found at http://www.ncbi.nlm.nih.gov/projects/GenBankUpdate/genbank_update.cgi.

3.6. Submitting a Sequence to GenBank® Using BankIt

The process of submitting data to the NCBI using the BankIt online submission form is as follows:

1. Login
   Login (or create a new user) at the PDA accessed at http://www.ncbi.nlm.nih.gov/WebSub/?tool=genbank and clicking on "sign in to use BankIt". After the login screen, there is the option to create a new submission or continue on a saved submission.

2. Contact
   The contact information of the person submitting the sequence to GenBank®. Subsequent BankIt submissions will retain this information and display it once the submitter logs in.

3. Reference
   The contact information of the contributing authors and information on publication that the sequence is referenced in (possibly tentative).

4. Nucleotide
   This tab contains information on when to release the submission for public access, either immediately or on a specified date.
   The definition lines are where information about the sequence is put in. The definition lines include the molecule type (generally genomic DNA for genomic authentication), topology (linear if it is a partial sequence) and genomic completeness (usually partial if it is a DNA barcode). The sequence itself is uploaded either by text input in FASTA format or by uploading a FASTA file. The sequence is input in the 5′→3′ direction. An example of how to fill this form out for the *E. arvense* matK sequence is presented in Fig. 9.

5. Submission category and type
   GenBank accepts both original sequences and annotations of already existing sequences. This section defines which of these two categories a submission fits into.

6. Source modifiers
   This section contains details on any specific details of the submission including organelle/location (such as plastid: chloroplast) and any additional source modifiers not included in the FASTA file.

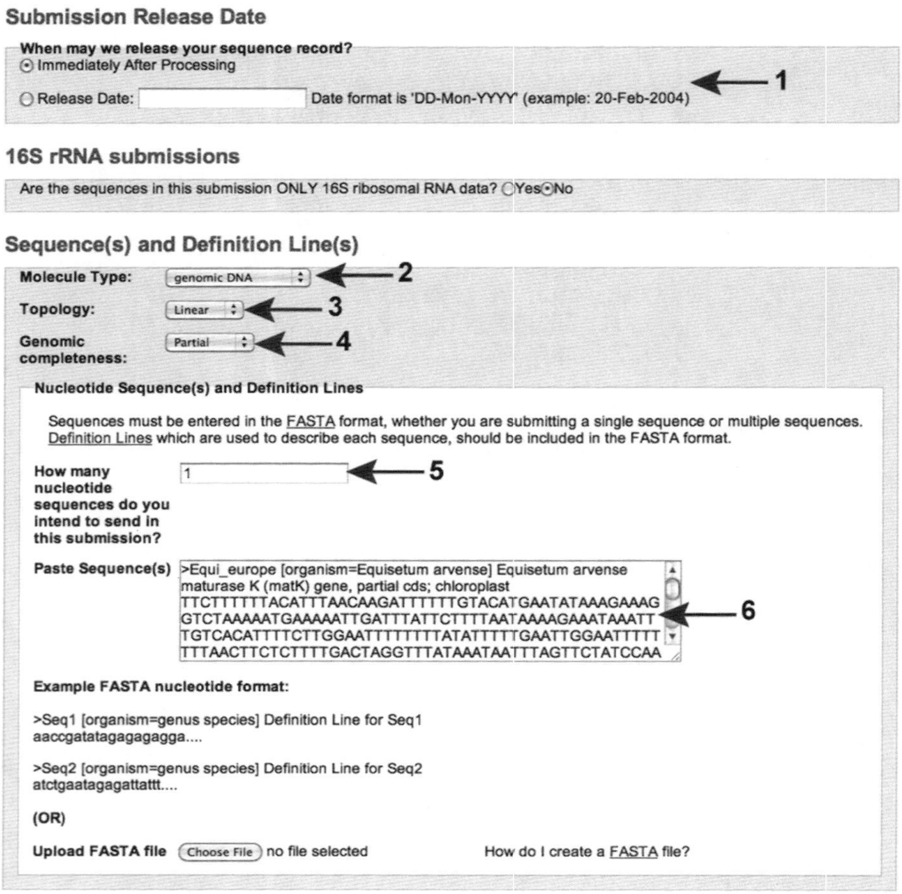

Submission Release Date

When may we release your sequence record?
⊙ Immediately After Processing

○ Release Date: [＿＿＿＿＿＿] Date format is 'DD-Mon-YYYY' (example: 20-Feb-2004) ◄━━**1**

16S rRNA submissions

Are the sequences in this submission ONLY 16S ribosomal RNA data? ○Yes⊙No

Sequence(s) and Definition Line(s)

Molecule Type: [genomic DNA   ⬍] ◄━━**2**

Topology: [Linear ⬍] ◄━━**3**

Genomic   [Partial ⬍] ◄━━**4**
completeness:

> **Nucleotide Sequence(s) and Definition Lines**
>
> Sequences must be entered in the FASTA format, whether you are submitting a single sequence or multiple sequences. Definition Lines which are used to describe each sequence, should be included in the FASTA format.
>
> How many   [1   ] ◄━━**5**
> nucleotide
> sequences do you
> intend to send in
> this submission?
>
> Paste Sequence(s) [>Equi_europe [organism=Equisetum arvense] Equisetum arvense
> maturase K (matK) gene, partial cds; chloroplast
> TTCTTTTTTACATTTAACAAGATTTTTTGTACATGAATATAAAGAAAG
> GTCTAAAAATGAAAAATTGATTTATTCTTTTAATAAAAGAAATAAATT ◄━━**6**
> TGTCACATTTTCTTGGAATTTTTTTTATATTTTTGAATTGGAATTTTT
> TTTAACTTCTCTTTTGACTAGGTTTATAAATAATTTAGTTCTATCCAA]
>
> **Example FASTA nucleotide format:**
>
> >Seq1 [organism=genus species] Definition Line for Seq1
> aaccgatatagagagagga....
>
> >Seq2 [organism=genus species] Definition Line for Seq2
> atctgaatagagattattt....
>
> **(OR)**
>
> **Upload FASTA file** (Choose File) no file selected   How do I create a FASTA file?

Fig. 9. A screenshot of the BankIt nucleotide form used to submit a *matK* sequence from *Equisetum arvense*. The options selected are: (1) Release the sequence immediately; (2) Genomic DNA; (3) Linear DNA; (4) Partial sequence; (5) Upload 1 sequence; (6) The sequence to upload in FASTA format (5′→3′ direction).

7. Primers
   This optional section details the primers used for the sequencing reaction detailing the primer sequence and name.

8. Features
   Features such as the gene and CDS can be added to a submission by either uploading a five-column feature table or by completing a form in BankIt. Details of how to make a five-column feature table are presented in Subheading "Tab-Delimited Five-Column Feature Table".

3.7. Submitting a Sequence to GenBank® Using Sequin

Sequin is a powerful tool rich in features. The tutorial that follows is designed to use only the very basic functions of Sequin to get a sequence submitted. A complete tutorial on Sequin is available from http://www.ncbi.nlm.nih.gov/Sequin/QuickGuide/sequin.htm.

The Sequin program can be downloaded at http://www.ncbi.nlm.
nih.gov/Sequin/download/seq_download.html.

1. Sequin welcome page
   Run the Sequin program and make sure the "database for
   submission" button is on "GenBank". Click "Start New
   Submission".

2. Submitting authors
   This form contains the details on when the submission should
   be released, the tentative title of the manuscript and the contact
   information of the contributing authors.

3. Sequence format
   This form is where the details of the particular sequence are
   input, such as whether it is an original submission or third
   party annotation.

4. Organism and sequences
   (a) Nucleotide tab
       This is where the actual sequence is input. It can either be
       loaded as a FASTA file using "import nucleotide FASTA"
       or directly using "add/modify sequences".

   (b) Organism tab
       This is where details of the organism the sequence came
       from are input.

   (c) Annotation
       This is where to add data about the region your sequence
       came from. You can add data such as the CDS and gene.
       Generally speaking, a sequence will not be a complete CDS
       and therefore the "incomplete at 5′ end" and "incomplete
       at 3′ end" should be ticked, otherwise an error may result
       during the translation.

5. The final page is a summary of the submission. Read through
   all the data carefully and, if necessary, edit any of the sections
   by double clicking on what needs to be edited.

6. When "done" is clicked, a list of detailed errors may appear.
   These need to be addressed before the sequence can be submitted to the NCBI, though the file can be saved without fix-
   ing them.
      The updated *.sqn file should then be emailed to gb-sub@
   ncbi.nlm.nih.gov or send directly using UpdateMacroSend
   (http://www.ncbi.nlm.nih.gov/projects/GenBankUpdate/
   genbank_update.cgi).

4. Notes

1. Nearly all of the information presented in this chapter has been retrieved from the NCBI Web site and condensed to form a concise tutorial. Detailed instructions can be found at the Web sites given for the relevant sections.

2. When using BLAST to identify a DNA barcode based on a database match, there is no absolute certainty that the match is actually that for the organism.

References

1. Luscombe NM, Greenbaum D, and Gerstein M (2001) What is Bioinformatics? Method Inform Med 40:346–358

2. Pool R and Esnayra J (2000) Bioinformatics: Converting Data to Knowledge: Workshop Summary. National Academy of Sciences, Washington D.C.

3. Benson DA, Karsch-Mizrachi I, Lipman DJ, et al (2008) GenBank. Nucleic Acids Res 36:25–30

4. National Center for Biotechnology Information (2007) BLAST Program Selection Guide. National Center for Biotechnology Information. http://www.ncbi.nlm.nih.gov/blast/product-table.shtml. Accessed 3 March 2011

5. Lipman DJ and Pearson WR (1985) Rapid and Sensitive Protein Similarity Searches. Science 227(4693):1435–1441

6. Madden T (2002) The BLAST Sequence Analysis Tool. In: McEntyre J & Ostell J (ed) The NCBI Handbook. National Center for Biotechnology Information, Maryland

7. Altschul SF, Gish W, Miller W, et al (1990) Basic Local Alignment Search Tool. J Mol Biol 215:403–410

8. Altschul SF (2011) The Statistics of Sequence Similarity Scores. National Center for Biotechnology Information. http://www.ncbi.nlm.nih.gov/BLAST/tutorial/Altschul-1.html. Accessed 18/03/2011

9. Zhang Z, Schwartz S, Wagner L, et al (2000) A Greedy Algorithm for Aligning DNA Sequences. J. Comput. Biol. 7(1–2):203–214

10. National Center for Biotechnology Information (2003) BankIt: GenBank Submissions by WWW. National Center for Biotechnology Information. http://www.ncbi.nlm.nih.gov/BankIt/help.html. Accessed 28 February 2011

11. National Center for Biotechnology Information (2010) Sequin application standard release. National Center for Biotechnology Information, Bethesda

12. National Center for Biotechnology Information (2007) Sequin for Database Submissions and Updates: A Quick Guide. National Center for Biotechnology Information. http://www.ncbi.nlm.nih.gov/Sequin/QuickGuide/sequin.htm. Accessed 20 March 2011

13. National Center for Biotechnology Information (2011) GenBank UpdateMacroSend. National Center for Biotechnology Information. http://www.ncbi.nlm.nih.gov/projects/GenBankUpdate/genbank_update.cgi. Accessed 20 March 2011

14. Tao T (2007) Explanation of HTML Output of a BLAST Search. National Center for Biotechnology Information. http://www.ncbi.nlm.nih.gov/staff/tao/URLAPI/new_view.html. Accessed 18/03/2011

15. National Center for Biotechnology Information (2010) Glossary. In: McEntyre J & Ostell J (ed) The NCBI Handbook. National Center for Biotechnology Information, Bethesda

16. National Center for Biotechnology Information (2011) Updating Information on GenBank Records. National Center for Biotechnology Information. http://www.ncbi.nlm.nih.gov/genbank/update.html. Accessed 20 March 2011

17. National Center for Biotechnology Information (2009) Submission of Annotation Using a Table. National Center for Biotechnology Information. http://www.ncbi.nlm.nih.gov/Sequin/table.html#Table%20Layout. Accessed 20 March 2011

18. National Center for Biotechnology Information (2010) Modifiers for FASTA Definition Lines. National Center for Biotechnology Information. http://www.ncbi.nlm.nih.gov/Sequin/modifiers.html. Accessed 18 March 2011

19. National Center for Biotechnology Information (2010) GenBank Feature Table Definition. National Center for Biotechnology Information. http://www.ncbi.nlm.nih.gov/projects/collab/FT/. Accessed 18 March 2011

INDEX

Nikolaus J. Sucher et al. (eds.), *Plant DNA Fingerprinting and Barcoding: Methods and Protocols*, Methods in Molecular Biology, vol. 862,
DOI 10.1007/978-1-61779-609-8, © Springer Science+Business Media, LLC 2012

Printed by Printforce, the Netherlands